THE WOODBURNERS HANDBOOK

Rekindling an old Romance

Dedicated to an old flame — Suzanne

THE WOODBURNERS HANDBOOK

David Havens

HARPSWELL PRESS
Brunswick, Maine

HARPSWELL PRESS
Simpsons Point Road
Brunswick, Maine 04011

International Standard Book Number; 0-88448-007-0
Library of Congress Catalog Card Number: 74-29362

Contents

FOREWORD...... 7

THE GRAB BAG: Purpose of Heated Environment......... 9
Considering Your Needs......10
Winterizing the House......11
Heat Loss Table......12

WOOD: Advantages as Fuel......13
Heating Value......13
Seasoning Wood......14
Moisture and Durability......16
Splitability......17
Felling the Tree......19
The Chain Saw......20
The Cord of Wood......22
Burning Wood......24

WOOD STOVES: A Brief History......26
Buying an Old Wood Stove......29
Old Wood Stove Engravings......33
New Wood Stoves......43
Accessories......47
Connecting Stove to Chimney......48
Using the Wood Stove......49

COOKING ON The Gist of It......51
WOOD STOVES: Boiling......52
Broiling and Baking......52

HOW TO BUILD SIMPLE AND PRACTICAL WOOD STOVES:

55 Gallon Drum Stove......................56
The Drum Stove Door......................57
Drum Stove with Heat Exchanger.......58
Drum Stove Lining............................60
Other Drum Stove Factors.................61
The Box Stove..................................63
The Brick Stove...............................66

CHIMNEYS: NEW, USED, AND PREFAB

Construction....................................67
Flue Size, Height, Support.................67
Flue Lining......................................69
Walls, Soot Pocket Cleanout..............70
Smoke Pipe Connection.....................72
Insulation..74
Roof Connection...............................75
Top Construction..............................76
Inspecting the Chimney.....................78
Repointing Brick Chimneys...............79
Lining a Linerless Chimney...............80
Prefabricated Metal and
Cinderblock Chimneys......................82
Cleaning the Chimney.......................84

AROUND, INSIDE, AND ON TOP OF FIREPLACES:

Design..86
Height, Footings, and Hearth............87
Walls, Jambs, and Lintel...................88
Throat and Damper...........................92
Smoke Shelf, Chamber, and Flue.......93
Modified Fireplaces..........................93

Foreword

The wind is dry and mean tonight, careening in madcap fashion around the corners of the house. The thermometer shows 28 degrees, and falling. A halo encircles the moon, bringing with it a promise of snow. And throughout the nation, fuel trucks come and go, making their rounds. Here and there they are a bit off schedule, and the price per gallon of what they leave behind is breathtaking. And a question that pops up with greater regularity each day rings loud and clear tonight: What does one do for heat if the trucks cease coming altogether?

Some people say that harnessing the power of wind is the answer. A network of stylized windmills stretched out across the Great Plains to supply us all with the bulk of our energy needs. That is a very sound and possible answer.

Not a few people believe that solar power is the answer. Indeed, a number of creative individuals have already employed it with much success on the home level.

Still others want to harness the tides, or convert methane gas into useable power.

All of those suggestions may one day come to be. As for now, they do not exist to a functional degree for whole populations. So, what will suffice as an alternative?

Enter onto the set the author of this book, David Havens. He is a man who took the problem by the horns and solved it. He went deep into the vaults of that curious structure known as the Cathedral of Nostalgia and came out with a very wonderful and simple solution: wood heat.

That is what this book is all about. It tells you how to select the right species to fit the right circumstances, how to dry it, split it, stack it, buy it, and so forth. This publication also introduces you to the wood stove. It tells you how they developed throughout history, how to buy and repair old stoves, how to buy the right new model, and how to make them work. For those who are cooks, there are tips on using a stove for that purpose.

This book also looks into fireplaces and down chimneys, to help you insure that they are safe and efficient. Should you want to build one or the other or both, the basic information is within.

If a method or technique proved to be easy and both safe and economical, it was included in the text. If it took a magician's talents and the backing of a covey of gods, it was excluded.

Upon completing the manuscript, David Havens hightailed it back to Wiscasset, Maine, where he lives with his wife, Suzanne, and their two children. I suspect that they are, at this writing, seated around the big woodstove in their kitchen, talking of things more pleasant than whether or not the fuel truck will make it in time to feed the hungry furnace. That problem no longer concerns him. His father's way is now his way. And I suspect it will always be so. For, the world may turn head-over-heels, but wood heat just goes on radiating its merry warmth.

Without further chatter, then, welcome to a most rewarding compilation of skills. What so nicely served those of earlier years is still very much applicable to today, energy crisis or not.

The Editor

The Grab Bag

Before running headlong into specifics, it may be wise to consider just what the wood stove demands of its user. It is a space heater, as opposed to a central heating system, and in order for it to operate at peak efficiency, the house must be in good structural order. And the user must be willing to make certain shifts in lifestyle, all of which are of a very practical nature.

The primary function of a heating system is to provide the body with an environment in which it can lose heat comfortably. When healthy, the body registers a constant temperature of 98.6 degrees Fahrenheit. The common temperature range within most houses lies between 68 and 78 degrees F. If such a temperature is found to be enjoyable, this means that the body finds it easy to expel its excess heat. If the temperature is too high, the body sweats to release that heat. If the thermometer plummets, the body resorts to shivering and shaking in order to maintain normalcy. It is the point in between those two extremes that is to be tried for. This desired compromise will differ between individuals, depending on eating, dressing and general living patterns.

The variety of wood stoves available for home heating take these differences into consideration. Airtight stoves are the most efficient woodburners and are used by those who need a constant source of heat. Open stoves are more aesthetic, but they deliver a much lower level of heat. Cast iron stoves hold heat for long periods of time but take longer to actually begin radiating their warmth. Sheet metal stoves radiate very quickly, but do not hold their warmth for long. Ceramic stoves hold heat for extremely long periods of time but must be of massive proportions to do so. Potbelly stoves are fine for those who wish to use both coal and wood, but they will not accept large logs and so demand greater attention. Box stoves burn wood only, take good-sized logs and will heat a large area very nicely. Some stoves come equipped with manual dampers and some have automatic dampers. The latter may mean more efficient heating in the long run, but they also break down with greater regularity.

It is clear, then, that if one lurches out the door to purchase just any kind of stove, dissatisfaction will arise in the near future. Forethought is a necessary ingredient. Perhaps a look at how I went about setting up my stoves will help.

My wife, two children and I have always seemed to gravitate towards the kitchen. It is the room we meet in before going off to school and jobs, and it is the room we return to at night. We read in it, talk in it, eat in it and so forth. Therefore, it gets continuous heat. The stove has a large firebox and radiates heat for long periods of time. It also has the ability to hold coals overnight for quick starting in the morning. Because the room is 12 x 12 feet, it is a compact stove. It also has a built-in oven and two cooking holes. This set-up works very well.

The next room in terms of importance is the family room, just off the kitchen. This is heated only at night, on an average of three days a week. It measures 17 x 24 feet and serves as a study, a sewing room, T.V. area, workshop, party room...whatever happens to be of special interest to the majority at the time. The stove in this room is a combination of sheet metal and cast iron, stands four feet tall, and is rounded in shape. It can take a log up to 28 inches long and 12 inches in diameter. The size of the stove deals with the space in the room very efficiently.

The dining room is usually very comfortable without the use of a stove, because it enjoys heat cast off from the kitchen and family room. If guests come to visit, we sometimes light up a fire in the fireplace. This is bolstered by using a Franklin fire frame which increases the heat radiation.

The bedrooms are seldom heated because floor radiators (grated holes in the floor of each bedroom) allow the warm air from the kitchen and dining room to rise. A third bedroom has provisions for a stove because it does not enjoy a location over the kitchen, but presently it is not in use.

Sleeping in unheated bedrooms was a shock for our family until we discovered bedwarmers. We now depend upon an antique soapstone and two modern bricks. These are heated in the stove oven throughout the day. Just before we get ready for bed, they are removed, wrapped in towels and slipped under the covers where they warm the bed. It is mighty pleasing to be able to reach your toes down toward the radiant heat throughout the night.

The living room, at the back of the house, is closed most of the time. It does contain a small parlor stove that can be fired up should the room be needed. From a temperature of 20 degrees above zero, the stove can heat the room to comfort in about one hour. It is very adequate for its intended use.

Having been raised in a house with central heating, I found that I had developed many useless in-house habits, not the least of which was wanderlust: moving from room to room with no particular purpose. Having a den, a pool room, a family room, a guest room, ad infinitum. The convenience of even temperatures throughout the house has since been replaced. When I up and move elsewhere in the house, it is now for good reason.

Once human needs have been determined, consider the house itself. A close scrutiny will no doubt reveal areas where heat trails to the great outdoors quick as a cat scats from a dog.

Perhaps in the name of convenience all the doors between rooms have been taken down. It's a wise idea to put them back up. They serve to trap heat. If there is a large gap between the house and the ground, it is a major source of heat loss. Sealing it off with tar paper or branches and leaves will turn it into dead air space, a good sort of insulation.

It may be necessary to paint the house and to caulk any cracks in the wood, siding or stone. If there is a chimney that's not intended for use, seal it up temporarily with a piece of tight-fitting plywood. Simply closing the flue will not always solve the problem of drafts as many flues fit too loosely. Apply weather stripping around the edges of the plywood.

Installing storm or double windows is a necessity. A single pane window that is exposed to a room temperature of 70 degrees and an outdoor temperature of 20 degrees will register a surface temperature of 32 degrees. What exists in that case is almost a literal sheet of ice facing the living area. If storm or double windows are employed, the glass surface temperature in the same situation will register 52 degrees, and that's quite a difference. Also, caulk around the inside window seams with rope putty. This is easily removed come spring. Hanging curtains over window areas is both decorative and economical in that it creates dead air space.

When checking the house for such obvious defects, be aware, too, of cool surfaces. Heat that comes in contact with cold surfaces is cooled instantly and then sinks to the floor, creating unpleasant drafts. Proper insulation in walls and ceilings will rectify that problem.

The table immediately following shows the reduction in heat loss attained by proper insulation:

3-5/8 inches insulation in sidewalls....20.3% heat loss reduction
3-5/8 inches insulation in ceilings.......14.5% heat loss reduction
Weather stripping on windows............ 9.8% heat loss reduction
Storm windows and doors..................31.3% heat loss reduction

TOTAL: 75.9% reduced heat loss

While improving upon natural defects, consider those installed in the name of modernity, such as Fido's "very own front door," through which he can exit whenever nature screams. Perhaps it would be best to seal off such openings and let him or her out in person. Always remember it is easier to conserve energy than to create new energy. The cost of insulation is a one-time expense, but heat which is lost to the outside costs money every day.

Wood

Wood has certain advantages as a fuel that many people fail to recognize, the most important being that it is available and can be gathered without complicated equipment.

Wood is clean and free from disagreeable dust. It produces little smoke or soot if burned properly. A cord of hardwood leaves only 60 pounds of ashes, while one ton of a fuel such as hard coal leaves behind 200 to 300 pounds. Wood ashes, moreover, have fertilizing value.

In relation to heating oil, wood is the more desirable fuel. Oil produces sulfur dioxide when it is burned and wood produces little, if any. Number 2 heating oil contains one-half of one per cent sulfur. Number 6 heating oil contains 2.5 per cent. Ninety per cent of the sulfur contained in oil is converted into sulfur dioxide when burned. Such emissions are hardly beneficial to living things.

Wood begins to burn at a comparatively low temperature and can be maintained at a lower ebb than other major fuels when only a small output of heat is needed. And finally, wood burned part of the time, when it is needed, is far cheaper than any other fuel burned all of the time.

HEATING VALUE OF WOOD

Generally, there are two kinds of wood, hard and soft. It is the former that is burned in stoves. Hardwood burns longer, more evenly, and can be controlled to a more efficient degree. Softwoods contain considerable quantities of resins and oils that tend to make them good fuel but also causes them to be consumed at a quicker rate. Pines, for example, make for a fast, hot fire and last a shorter time than birch, but birch gives a more intense flame than oak. Oak and hickory burn more slowly and give an even heat.

American beech has long been a favorite fuel wood in the northeastern and central regions of the country, having heating value nearly equal to that of the best oaks. Eastern hophornbeam, better known as ironwood, is also very heavy and yields much heat per cord. Red mulberry and hawthorn, though small trees, are highly esteemed as fuel. Of about the same fuel value as sycamore is black tupelo (known as "sour gum" and "black gum"), which is widely distributed in the Eastern States. For use in open stoves and fireplaces, chestnut, butternut, tamarack, and spruce are not generally in favor, because they throw off sparks.

Moisture in the wood is the most important factor in the heating value. When wood is burned, the water in it must be raised to the boiling point, converted into steam, and finally superheated to the temperature of the flue gases. Hence, the heat required to drive off the moisture does not serve to warm the stove or furnace. Generally from 25 to 45 per cent of the weight of green wood is water, and in such species as cottonwood and willow it may be even 55 or 60 per cent. By drying out much of this water, the heating value of the wood is increased considerably. For example, green shagbark hickory weighs about 800 pounds more per cord than air-dried wood of the same species, the difference representing mostly water. This extra moisture reduces the heat value about one-sixth, as shown in the table on the following page.

Drying the wood for a short time is better than not drying it at all; if air is allowed to circulate freely about the wood for three months in reasonably dry weather, seasoning will be about half complete, and the fuel value will then be about 90 per cent of that of thoroughly air-dried wood. Dry wood kindles more readily than wet wood. In a stove a fire of dry wood is generally easier to tend and regulate. It is said, however, that some species such as gray birch and aspen give better results if not too dry, being consumed less rapidly.

SEASONING WOOD

For green freshly cut from the forest, from 6 months to a year is usually required for thorough seasoning. The rate of evaporation of the moisture will depend on such factors as temperatures, relative humidity in the air, exposure to rain or snow, and movement of the air about the individual sticks. Splitting the pieces helps and is

Approximate weight and heating value per cord of different woods,
green and air-dry

Species	Weight		Available heat	
	Green	Air-dry	Green	Air-dry
			Million BTU	Million BTU
	Pounds	Pounds		
Ash..............................	3,840	3,440	16.5	20.0
Aspen..........................	3,440	2,160	10.3	12.5
Beech, American........	4,320	3,760	17.3	21.8
Birch, yellow..............	4,560	3,680	17.3	21.3
Elm, American...........	4,320	2,960	14.3	17.2
Hickory, shagbark......	5,040	4,240	20.7	24.6
Maple, red..................	4,000	3,200	15.0	18.6
Maple, sugar..............	4,480	3,680	18.4	21.3
Oak, red.....................	5,120	3,680	17.9	21.3
Oak, white.................	5,040	3,920	19.2	22.7
Pine, eastern white.....	2,880	2,080	12.1	13.3

Air-dry means with 20 per cent moisture in terms of oven-dry weight, or 16.7 per cent in terms of total air-dry weight. One BTU (British thermal unit) is the amount of heat required to raise the temperature of one pound of water 1° F. Available heat equals calorific value, minus loss due to moisture, minus loss due to water vapor formed, minus loss in heat carried away in dry chimney gas. Flue temperature 450° F.; no excess air. (Data supplied by Forest Products Laboratory, Madison, Wis.)

almost indispensable with some species such as birch and alder, which should be placed in fairly dry locations. Moreover, cottonwood for example, is easier to split when green.

The wood should be stacked outdoors where it is exposed to sun and wind, and preferably on a hilltop. Skids may be used to keep the pieces off the ground, and for the best results either some form of roof or covering should be provided. The top layer of sticks can be packed closely and slanted, to help prevent rain from reaching the interior of the pile. In order to accelerate the seasoning process, the sticks may be piled in crisscross fashion. This provides for maximum circulation of air. If branches of live trees felled during the summer are left intact for 2 or 3 weeks, considerable moisture will be drawn out through the leaves.

Tree species vary considerably in the proportions of moisture usually contained in their wood at the time of felling. For example, green Douglas-fir heartwood is so dry that it has very little extra fuel value when seasoned. Near the opposite extremes are the cottonwoods, which contain a great deal of water. Under emergency situations, it may be desirable to cut fuel wood for use within a very short time. In such circumstances, if the available woodland offers a choice of several species, it would be useful to know the kinds of trees from which satisfactory fuel may be produced with a bare minimum of seasoning. The following species are improved comparatively little by drying:

Ash, biltmore	Hickory, pignut
Ash, blue	Hickory, shagbark
Ash, Oregon	Locust, black
Ash, white	Osageorange
Beech, American	Pine, Lodgepole
Douglas-fir	Pine, red
Fir, alpine	Spruce, red
Fir, noble	Spruce, white

Higher moisture content makes seasoning of the following woods more necessary:

Alder, red	Hackberry	Pine, loblolly
Ash, black	Hickory, bitternut	Pine, pitch
Birch, paper	Hickory, nutmeg	Pine, ponderosa
Birch, river	Hickory, water (Bitter pecan)	Pine, shortleaf
Cottonwoods	Honeylocust	Pine, sugar
Elm, American	Maples	Sugarberry
Fir, grand	Oaks	Sweetgum (redgum)
Fir, Pacific silver	Pine, jack	Tupelo, water
Fir, white	Pine, jeffrey	Walnut, black

Once the wood is seasoned and sits stockpiled in the consumer's yard or shed, another important factor enters into the picture: durability. Durability has nothing to do with the hardness or softness of the wood itself, but depends upon a certain balance of resins and tannins. Some woods are more susceptible to rot and fungi than others if exposed to temperatures of 60 to 90 degrees, moisture and sufficient oxygen. Eliminate one of those three and fungi cannot exist. In order to further guard against decay, the following list will enable selection of the most durable woods available in a given area:

VERY DURABLE	DURABLE	INTERMEDIATE	PERISHABLE
Black Locust	White Oak	White Pine	White Elm
Eastern Red Cedar	Black Ash	Norway Pine	Beech
Live Oak	Cherry	Shortleaf Pine	Hickory
Black Walnut	Red Elm	Red Oak	Hard Maple
Cypress	Persimmon	Red Ash	Red Gum
Western Red Cedar	Longleaf Pine	Yellow Poplar	White Ash
Redwood	Western Larch	Butternut	Loblolly Pine
White Cedar	Slash Pine	Sugar Pine	Hemlock
	Ironwood	Sugar Pine	Spruce
			Yellow Birch

SPLITABILITY

Another factor to consider when choosing wood is its "splitability." What looks so very easy in the rough and ready motion picture may prove to be somewhat frustrating in real life.

The least line of resistance on a piece of wood is found along the radius, from which point wood rays emanate. The straighter the grain, the easier the task will be. Wood with wavey or twisting grains, such as elm, will prove to be difficult. When met with such a piece, toss it aside for use as a yule log. The same is true of knotty wood. Never have more epithets blared across the countryside than when someone attempts to split such a piece.

Always chop the log on top of another wider log or stump. If chopped directly on the ground, the soft earth will absorb most of the shock and the return for effort will be halved.

The list on the following page should be of assistance in gauging the splitability of some major woods.

HARD	INTERMEDIATE	EASY
Elm	Birch	Chestnut
Beech	Maple	All Pines
Black Gum	Hickory	Redwood
Sycamore	Red Oak	Cedar
Dogwood	Ash	Fir
Red Gum	Cottonwood	Western Larch

As for the tools necessary to do the job, the axe seems to be most efficient. Some prefer the splitting mall or a sledge hammer and a steel wedge with which to force the wood apart. The axe, once one gets the hang of it, accomplishes the same task with one sharp blow. However, keep a splitting mall and a few wedges handy in case a comparatively hard piece shows itself and will only succumb to brute force.

Aside from the major stockpile, kindling is a necessity. It should be small and very dry. Some say that pine cones work very well, as they contain a great deal of sap which burns quickly and radiates a

great amount of heat. Corncobs and dried citrus peels also work well. Both contain natural oils that combust rapidly. All softwoods, especially pine, will serve well. If the wood is green, it will have to sit for two years before it is dry enough. Kindling will mean scrounging the forest floor or the lumber yard for the butt ends of two-by-fours, or any wood factory offering scrap wood. Some people use lobster pot slats. The key to good kindling, aside from its dryness, is that it be very small. One can save on kindling if the stove coals are kept hot constantly.

FELLING THE TREE

Eventually, someone is going to get the urge to cut down his or her own tree. It happens to the best of us, and since there is a shortage of hospital beds around the country, some advice is called for.

The process of felling a tree should really not be attempted without first gaining the benefit of instruction on the spot by someone who has done it before.

The basic tools needed to do the job safely are: block and tackle, a pulley, chain saw or axe, a sledge hammer and wedges, and a ladder.

Once the tree has been selected, the next step is to determine the direction of fall. If in the woods and away from overhanging wires, our major concern will be overhanging and surrounding trees that might interrupt or stop completely the fall of the tree. It's wise to clear a path for it. This can be approached with a certain amount of modesty, as there is no need to ravage the surrounding forest just to bring down one lone member in it.

The next step is to make a fairly deep wedge-shape cut at the base of the tree, on the side pointing in the fall direction. Then attach the block and tackle to the tree at a high elevation and, if no help is available, run it to another tree in front and to the side of it. Keep the block and tackle taut at all times during cutting. It is your major assurance that the tree will not fall backwards.

The actual felling of the tree is accomplished by going to the opposite side from the initial wedge-shaped cut and working towards the fall direction. As the cut grows deeper, the tree may tend to jerk back some, thus pinching the blade of the saw or axe. This can be

A birch log showing growth rings, radius marks and heart wood. Finger is pointing to heart wood.

corrected by inserting a wedge in the cut. Eventually, the tree will fall exactly where it was supposed to.

Once it is down, cut off all the branches. When the trunk is bare, cut it up into moveable lengths. Lumbermen usually cut it into cord lengths, but it may be better to cut it into stove lengths on the spot. This means more carrying to and fro, but the pieces are easier to deal with if an average car or pick-up truck is being used to haul the wood. It also gets one in the frame of mind of using every available twig.

As the trunk is being cut, the problem of pinching may again arise. The best advice is to support the horizontal trunk so that it cannot sag. Use whatever is on the site, or better yet, bring such support from home...and take it back.

The chain saw, if handled carelessly, can be a lethal instrument. Always keep the saw in proper working order. Most chain saws are two-cycle engines, which means gas and oil must be mixed as engine fuel. Every manufacturer recommends different proportions and kinds of oil. Follow these to the letter. Never use anything but regular gas.

Oil is also used to lubricate the chain. Chain-lubricating oil is different from the usual oils. It has a special sticky ingredient which helps it stick to the chain. Do not use regular oil as it will cause chain and bar wear.

Points and plugs should always be kept clean and properly gapped. Check manufacturers' specifications. The easiest way to check the spark is to remove the plug, attach the wire as if in normal running situation, ground the plug to the engine and pull the starter cord. This should produce a nice blue spark. If the spark is weak, adjust or replace points or plug.

Chain and bar maintenance is just as important as engine maintenance. Always keep the drive mechanism clear of dirt, the chain adjusted properly and the teeth sharp.

To maintain the teeth in good condition, keep the saw out of dirt and learn to use the rattail file. Never use the saw if the teeth are dull. This will cause overwork on the engine and overheating on the chain and bar. Stop cutting every half hour and check the teeth for nicks or dullness. A quick pass with the rattail file over

A chain saw cutting wood for a stove. Note the correct hand positions.

the cutting faces will maintain the edge. Learn the proper angle for the cutting face and how to use the rattail file in maintaining this angle. If the user prefers, most chain saw shops can sharpen the saw with special machines which do an excellent job.

The best advice is to spend time talking to the man who sells the saw. Most dealers are proud of their products and will take time to instruct you in proper handling. When buying a chain saw, don't look for the cheapest deal. Ask around and discover who offers the best maintenance. Long-term maintenance will prove more important than original price.

THE CORD

One full cord of wood equals a stack measuring 4 x 4 x 8 feet. Five cords of wood for every two rooms to be heated during a winter is sufficient. A few modern heating specialists may differ with that formula, but it has withstood the test of many New England generations. Never has a man or woman spent a chilly night or day adhering to it.

Buying wood is, at best, a difficult proposition. Wood is usually sold by the cord or by the face cord, which is half a cord; be sure the seller makes clear which is which. The most expensive wood is aged hardwood. Most people prefer two-year-old wood for opti-

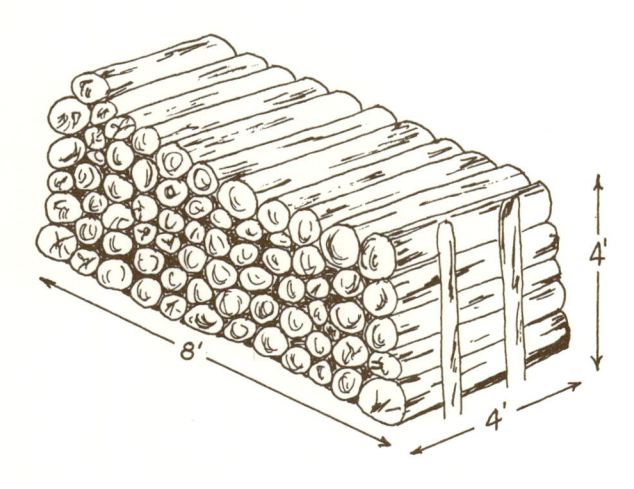

Mr. Richard Verney of Wiscasset, Maine holding the tools of a professional cordwoodsman. An antique splitter, and a modern gasoline chain saw. Note the size of splitting blades.

mum heat. Next to the kind of wood and its age or dryness is its splitting and cutting condition. Wood prepared for cooking stoves requires more work than wood split into large pieces. All these factors deserve consideration in the cost of a cord of stove wood. The buyer should be careful to specify exactly what he wants and be sure to determine exact measurements of wood received.

Most hardwood will contain at least fifty percent moisture when cut. After only six months, this figure will be reduced to twelve per cent if the wood is stored under cover and where air can easily circulate.

Some people ask why wood is not sold as coal used to be sold: by the ton. That would be a fair method of sale if all wood sold was equally dry, but that is not the case.

The types of cords available can be classified as follows: hardwood or softwood, well-seasoned, partly-seasoned or green; slabwood, dry or green; mill ends (softwood scraps). Be specific as to which type of cord you want, and the dealer will be better able to arrive at a fair and honest price.

A cord-cutting saw. Wood is sitting on the rack. This saw enables a good man to cut a cord of wood very fast.

The old folks in New England estimate that every wood-burning homeowner needs a fifteen-acre woodlot. This they contend will supply all the heating needs for the average family for the rest of their lives and on forever. Environmentalists in New England estimate that an acre woodlot can supply a cord of wood every year without any ecological damage.

HOW TO BURN WOOD

When wood is used as fuel, certain rules should be applied in adapting and operating the heating equipment. This will be beneficial in getting as much heating value from the wood as possible and will make its use generally more satisfactory.

1) Stove wood must be cut short enough to lie flat in the firebox. If the firebox is rounded or oval in shape, it is best to make the pieces somewhat shorter than its inside length. Several sticks

should be packed in closely, side by side, with only very narrow spaces between them, if a hot fire is wanted. The heat reflected from one piece to another helps drive off moisture and maintains the proper rate of burning.

2) Because wood burns with a long flame and intense heat, various measures should be taken to keep flames from going up the pipe and wasting the heat. Fuel should not be piled up near the level of the smoke outlet. With dry wood, the problem is to hold down the rate of release of combustible gas from the wood and at the same time to admit enough oxygen to burn all the gas that is released. The draft from below the grate on a wood stove should therefore be restricted carefully, while about four-fifths of the air is admitted around and above the fuel, usually through slots in the fuel door, or by way of draft valves. For the sake of efficiency, a check damper in the smoke pipe should be used. Wood does not need as much chimney draft as coal.

3) Holding a fire overnight requires extra fueling with the largest chunks, preferably of heavy hardwood, and special attention to closing the draft dampers tightly. Most stoves can be banked to hold a fire for ten to twelve hours without any attention.

4) In stoves, green wood should not be used in a slow fire (such as might be needed for certain types of cooking) because it will tend to cause soot deposits, creosote, and acetic acid in the smoke pipe and flue. Burn the green wood in a hot fire and save well-seasoned sticks for use in starting the fire or maintaining it at a low rate. A small, hot fire is better than a large, smoldering one.

Wood Stoves

The wood stove dates back in history well beyond the founding of America. In fact, the oldest known cast iron object is a stove which was made in China between 25 A.D. and 200 A.D. Now residing in the Chicago Museum of Natural History, it is horseshoe shaped, with a chimney at the rounded end and a platform in front of the firebox. It also has five cooking holes and legs shaped like an elephant's feet.

The first records of cast iron stoves in Europe date around 1475 and were found in Alsace. The stove was also in use in Germany by the year 1509 and in the British Isles by the mid-Eighteenth Century. Records in the Carron Iron Works of Scotland indicate that stoves were being cast in 1759 by the foundry.

In America, the first recorded stoves were made of wood and lined with clay. A patent for a cast-iron stove was issued by the General Court of the Massachusetts Bay Colony in 1652 to one John Clark. The earliest cast-iron stoves were made by the Saugus Iron Works in Saugus, Massachusetts in 1674.

The first European stoves were fashioned out of cast iron from five rectangular plates which formed a box with an open front. The open end of this stove was usually placed facing one room while the box itself stood in another. The room containing the bulk of the box (which rested on two legs) received the heat, while the other room received the smoke. There was no fuel door, no draft door, and no chimney. Just an open end through which the wood was fed, the air entered, and the smoke exited. This type of stove was known as the "Jamb," "Five-plate" or "Non-ventilating" stove. It was commonly used in Germany and other Scandinavian countries, with the exception of Holland.

In 1741, the manufacture of these stoves began in Pennsylvania using wooden molds imported from Germany. Decorations were usually based on Biblical themes and often had German words incorporated into the design. For this reason, they were called

"German Stoves." There are several on display at the Bucks County Historical Society Museum at Doylestown, Pennsylvania.

During the same period, the six-plate stove was in use in Holland. It was referred to as the "Draft," "Wind" and "Holland." This stove included major developments that earned it the nickname, "Grandfather of Wood Stoves." Cast of six separate plates, the front plate had a door and draft hole. The top plate contained a hole for a smoke pipe. The stove was designed to rest entirely in one room and radiated heat with surprising efficiency.

The New England version of the six-plate stove had three or four legs averaging 12 to 14 inches in height. Later developments included tie rods connecting the top, bottom and side plates. Some of the original models had heat exchangers which gave the appearance of a series of organ pipes. The smoke was forced to go through a lengthy "up and down" route before it was expelled into the chimney. Another variation of this heat exchange method existed in the form of a series of boxes on top of the firebox.

Around 1840, two vertical flues instead of one became popular in America. These were connected by a horizontal pipe that had a T-joint leading to the chimney. They were referred to, accurately enough as "Two Column Stoves."

In the 1820's a patent was granted for the development of a sunken pit for ashes. This was a moveable pan that fit under the firebox grate and was easily removed for ash disposal.

The boiling hole for the ever-present tea kettle was the next major advancement. Between 1843 and 1853, literally hundreds of design patents were granted. During this period stoves were decorated with fruit and floral designs. The Greek revival soon followed. Grecian forms and urns were predominant. These decorations are extremely helpful to the antique stove enthusiast for dating and casting.

Many stoves took their names from the functions they served, such as the "Four O'clock Stove." This was small and burned charcoal to produce sufficient heat for bedrooms. It was lit around 4 p.m. to take the chill out of the air before retiring. The "Little Cod" was used on fishing vessels and was so named because of the cod fish designs on its sides. The "Jews Harp" looked very much like its namesake.

Although not a stove in the strict sense, the Franklin Fireplace was immensely popular with Americans. Ben Franklin developed it because he wanted the efficiency of a stove combined with the aesthetics of a fireplace. Early models were made of flat castings and only later was the development of curved sides introduced. They served to radiate the heat more evenly about the room.

The first Franklins were decorated with a sun design showing sixteen rays. This was surrounded with branching leafing and streamers bearing the Latin inscription, "Alter Idem," or, "Another Like Me," testimony to their popularity. After the American Revolution, the stoves bore ornaments depicting Washington and Franklin. Later, the eagle and stars were introduced.

A peculiar adaption of the Franklin stove appeared in New England around 1800. It was made of three castings which could be fitted into a fireplace and then bricked in place. This added to the potential heat radiation of the fireplace and was, of course, cheaper than an actual Franklin. It is referred to as a "fire frame."

The first round stove dates back to 15th century Germany and was constructed from sheets of metal riveted together to produce its cylindrical shape. Others were cast of one piece and were known as "cannon" stoves. Such stoves were used as early as 1752 in New York City and were advertised in the *Boston News Letter*, dated February 1, 1770.

The first airtight stove was invented by Isaac Orr of Washington, D.C. in 1836. It had no seams and was oval in shape. Made of sheet iron, it had a door for front loading and a door in the top. The lower portion of the stove was lined with a heavier piece of iron. Standing on four-inch legs, it was said to have burned any solid fuel but coal.

In 1842, a man by the name of Elisha Foote of Geneva, New York invented a regulating draft which consisted of two bimetallic rods that could be adjusted by a setscrew. This was first used on box stoves, but in 1849 was adapted to fit airtight models.

The soapstone stove, developed around 1797, was the first such stove type to be built in America. Made from stone taken at the Francestown quarries in New Hampshire, it had the advantage of retaining heat better than cast iron. A number of Francestown homes still have functioning soapstone stoves.

Brick, earthenware and tile stoves were commonly used throughout European countries which experienced severe winters. This type of stove produces an even, comfortable warmth that lasts a long while after the fire has gone out. The brick models, known as "Copenhagens," were made in the early 19th century. A number of them were built in Salem, Massachusetts.

The first functional cook stoves were available in 1765, but housewives were reluctant to switch from the security of brick oven fireplace ranges. The first cook stove to enjoy widespread popularity in the United States was the "James" stove, first patented in 1815. Seven years later it included a sunken hearth, which has been a regular feature on most cook stoves ever since.

The Conant stove of 1819 improved on the "James" stove by adding a lowered pot hole which resulted in faster heating of contents. It also included an oven over the fire with a door at both ends.

The "Premium" stove was a square box containing an open grate and a high ash pit, to which was later added an elevated oven, located to the side of the firebox. In 1838, the ash pit was improved to lie at a shallower depth.

Gas recycling stoves were introduced in 1682 with a model called the "smoke consuming stove." In 1771, Ben Franklin developed a smokeless stove shaped like a vase. It had an internal tube that sucked the smoke down into the burning coals. The smoke was then rekindled, as it were. Many companies took hold of the basic idea and developed home heating stoves with extremely complicated drafting systems. These recycling principles are now basic on all sophisticated heating furnaces, but are very rarely incorporated into modern wood stoves.

Many innovations put forth in yesteryears have been forgotten because most people have lost interest in the stove as a primary heating and cooking device. But good models that reflect the sensibilities of their forebears do exist.

BUYING AN OLD STOVE

Many people have a fascination for old stoves, either because they are handsomely designed, because they are strong and efficient, or a combination of both. Purchasing such a stove should

should be approached with a certain amount of caution and the folling tips may be helpful.

The most important features on any stove are the firebox and the grates. If one of the grates is missing or damaged, ask around for a replacement. If none exists on the open market, go to a foundry and request that they search their inventory for the grate designed to fit the stove. The foundry people will surface with either the grate itself or its pattern. The latter means casting a new piece. Sometimes neither grate nor pattern exist, in which case a new mold must be carved out of wood and cast. This can lead to some expense, but if the stove is desired, it is worth doing. Home repairs of grates can be accomplished, but all too often the heat melts the makeshift item and causes it to warp beyond use.

If the firebox is cracked to an extreme degree, it is wise to pass on to another stove. If the cracks are minimal, they can be repaired through welding. Such repair is difficult, as the weld must be even on both sides of the crack. The finished product should be a smooth, rounded surface with small and consistent ridges. Unless one has had the experience in welding techniques, it would be better to turn the job over to a professional. The firebox is a most critical area and must be in top shape at all times.

If the stove has a complicated draft system, be sure that all the parts exist and that none have been welded or rusted shut. It helps, too, to purchase a model on which the draft valves are within easy reach for regulation and safety checks. If the drafts are near the firebox and have been destroyed, new ones can be fabricated. Rusted levers can be loosened with Liquid Wrench.

Rust and pitting on the surface of a stove can be eliminated by using stove polish which is available at most hardware stores. Be sure that the rust has not eaten its way into the firebox or any other portion of the stove necessary to its functioning well. If rust has corroded an essential part, the best advice is to have it recast at a foundry. If there is sheet metal on the stove, it is easily replaced. Take the part to a sheet metal factory and they'll construct a duplicate. If the part has corroded beyond recognition, a good sheet metal person will most always be able to fashion a precise duplicate based on dimensions alone.

Check all doors and hinges. Many models were poorly cast and the hinges break. If the doors are warped, they will need to be replaced. It is impossible to straighten them out, and they are a

A rusted old cook stove put outside to die. This could be shiny and efficient with a little work. Note the front lever for the oven door.

crucial feature for efficient wood burning to take place. Hinge pins and most bolts can easily be replaced if broken or missing. New doors can be fashioned from heavy sheet metal if cast iron replacements are impossible.

Be sure that the stove in mind will burn what you think it will. A small firebox may indicate the use of coal only. And a stove designed to heat one room will never heat three adequately.

If the general body of the stove has a few small cracks in it, and they do not affect burning efficiency, they can be sealed with Rutland Stove Cement. Cracks of a greater nature can be repaired by smoothing down the rough edges, inside and out. Then drill a series of holes down the sides of the crack. Bolt heavy gauge steel plates over the crack, inside and out. To insure a tight seal, fill the plate seams in with Stove Cement, or have them welded.

I don't mean to discourage all home repairs. In fact, I've saved myself money by doing my own stove work. But familiarity with the stove as a working apparatus is necessary before such repairs are undertaken, or the balance of the whole may be destroyed. My father, who grew up in a home heated with wood, sent the following letter from the Belgian Congo to his mother. It is dated September 15, 1925 and perhaps serves as testimony to a bit of ingenuity in regard to stove repair:

Dear Folks at Home,

I've heard it said that if anything could get a preacher into the notion of saying "bad words," it is putting up a stove. Well, I "hain't" said none, but I can see how a fellow might. I set in this morning to erect a stove in our cookhouse. The stove had been abandoned and had been standing in the damp Congo weather for a long while. It is just a little more than a pile of rust.

She only had three legs, so I corrected that with a stack of bricks. But during the operation, something slipped and she suddenly threw herself at our feet on the floor, with two more legs broken square off. We elevated her slowly and more gently to the height of three stocks of bricks and one leg. We named her "Peg."

I looked around the sides and found them not so good. I had to hold up one end of the grate with a brick embedded in fire clay, and with the same material made a new lining for the stomach, as it was missing entirely. As for the oven, the top fell down and it was too rusty to replace, so we had no oven, only an extended firebox. This may sound as though it isn't a very good stove, but it is. The extra religion one gets working with her is a great compensation...

Mary Sue reminds me to tell you of our trouble at finding stovepipe. We resurrected three pipe joints but no "L's" could be found. So I took an old five gallon kerosene can, cut a hole in the bottom and one in the side, and it works as good as any "L".....

A smoke recycling stove with a built-in oven, teakettle attachment, anti-clinker door, elaborate styling and nickel foot rails. This stove is the ultimate in heating efficiency.

This handsome parlor stove was advertised as having a "Russia body," which was the early term for sheet metal.

The Station Agent, for offices, stores, school houses, railroad stations, chapels, etc. This stove is very strong and durable. It has a heavy cast pot, projecting umbrella top, shaking and draw center grate, nickel trimmings with a foot rail.

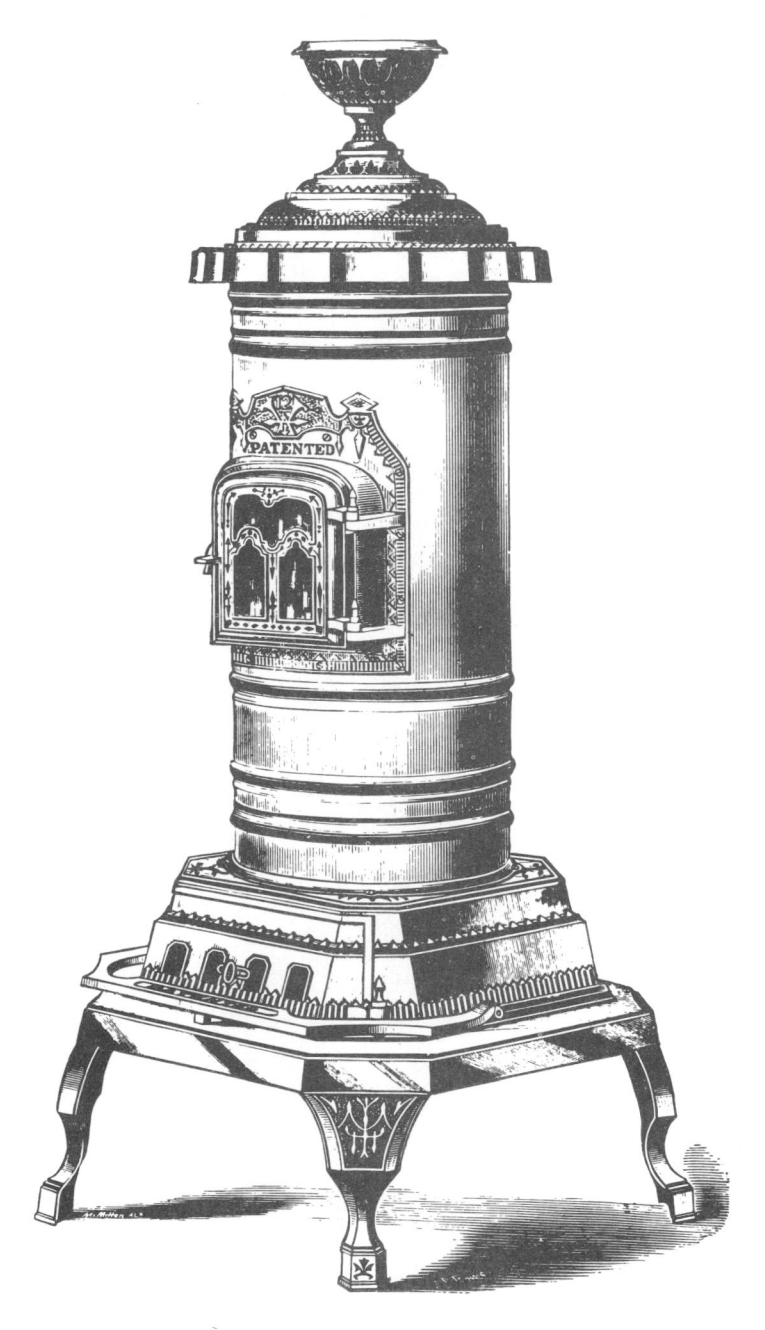

A round sheet-metal parlor stove with an anti-clinker door made to burn coal.

An artistic wood parlor stove with top, side, and front loading doors for ease in wood loading. Notice the elaborate bas-relief, nickel foot rail and trimmings which were typical of parlor stoves.

A heavy-duty parlor stove with front, top and side loading. The ornament on the top was often used to hold perfumed water which scented the entire room when heated.

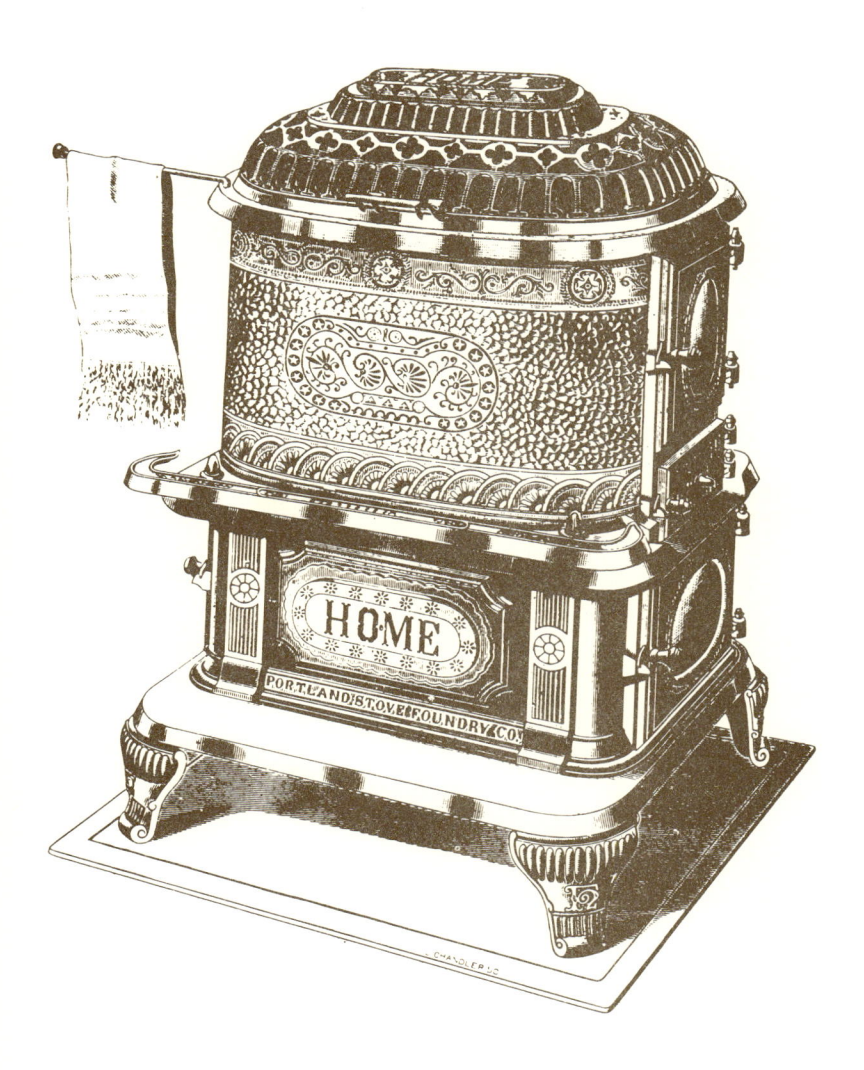

The Home Parlor Cook was one of the most popular stoves because of its versatility. It has an oven below the firebox with two doors and a drying rack. It burns wood or coal. The stove heats, boils, bakes or broils. It has a large firebox that will burn for 12 hours. The firebox is 22" long and the oven is 23" x 10" x 9".

This cooking stove or "range," as it used to be called, has a mantel oven, copper water tank and the usual oven beside the firebox. Special dampers allow the heat from the firebox to be concentrated on either of the ovens or the water tank. It also has drying racks next to the firebox and a vent for broiling over the fire. It burns wood or coal.

This stove has a temperature gauge on the oven which is not calibrated in degrees but has numbers one through five. The stove does not have a water tank.

This is the ultimate in home cook stoves. It has warming ovens, drying racks, water tanks, water coils around the firebox, front and side feeding doors for the firebox, a multitude of draft controls and a foot pedal for opening the oven door.

NEW STOVES

Many excellent stoves are now being produced, but the buyer must be able to match the capability of the stove with his own needs. Be wary in the new stove market, however, of the poorly designed and manufactured product. I have recently seen sheet metal stoves being sold which have no draft systems and wouldn't be able to maintain a fire. Even more frightening are stoves which are fire hazards. If you are new to stoves, stick with the ones produced by reliable manufacturers.

Since stoves come in all sizes, shapes, colors and materials, the following brief classification system may help. I classify stoves according to the kinds of draft systems, shapes, materials, uses, and where they are manufactured.

DRAFT SYSTEMS

Automatic Airtight Stoves:

There are many companies all over the world producing automatic airtight stoves. These stoves are designed with automatic thermostats which control the air intake to the firebox. In this manner, the fire can be smothered or fed on demand of the thermostat. The advantage of this feature is that the fire can be maintained for long periods.

Stoves with automatic thermostats can often keep fires going for 24 hours. The only limiting factors are the condition of the wood and the size of the firebox. Stoves with this feature are designed for full-time heating systems.

Manual Airtight Stoves:

Most manual airtight stoves are made from sheet metal and are inexpensive. Usually designed and produced for the vacation or emergency market, they are very efficient wood consumers, but are not designed to last forever. Be sure to follow the manufacturer's recommendations of putting sand in the bottom of the firebox.

Open Stoves:

Open stoves have less firebox surface because the front is open. Though they do not even begin to match the efficiency of a closed stove, they are one step better than a fireplace. A few companies offer models that have accordion doors on them that can be closed or opened, depending on one's whim. Some models allow for doors to be added to them. However, very few of the open stoves offering such door additions can be classified as closed stoves. The doors simply don't shut tight. There is only one variety of stove (that I am familiar with) that can be closed tightly enough with its optional door to be classified as a closed stove. It is a Norwegian model produced by Jøtul Stoves, Inc., of Oslo, Norway. It has a fire-viewing space as large as most open stoves and the workmanship is superior to all.

Closed Stoves:

The closed stove is a compromise between an airtight stove and an open stove. It is not airtight, so automatic thermostats cannot be used, but it is more efficient than an open stove. It usually has a front or side door which can be opened for viewing the fire, but it works best when closed. Closed stoves come in all sizes, shapes, colors and materials. The closed stove is a happy compromise for many.

STOVE SHAPES

Stove shapes can be classified as either round or square.

The most popular type of round stove is the Potbelly. Railroads gave them widespread exposure by using them in railway cars and stations. These are good, efficient, closed systems because their rounded shape packs a lot of surface into a small space. They come in a variety of sizes, shapes, materials and colors, and most are equipped with a pothole for cooking or warming teakettles. Potbellies are designed to burn coal, coke, or short wood. They are also equipped with grates and draft systems that feed air to the fire from below the firebox. The disadvantage of a round stove is that quite short pieces of wood must be used.

Box stoves are specifically designed to burn only wood. Their shape allows whole logs to be used, often without splitting. Some of these stoves are designed with cooking pots and front loading doors. Since there are no grates, no other material can be burned except wood. Box stoves are usually made from cast iron, though recently many sheet-metal ones have appeared on the market.

MATERIAL USED IN MAKING STOVES

This is an important consideration for the stove buyer in that the material, as well as the shape and design, determines the characteristics of the stove.

The most common materials are sheet metal, cast iron, bricks and tile. Each of these has different heat transference characteristics. Sheet metal transfers heat quickly by radiating all the energy created by the combustion in the firebox. This means that a stove made of sheet metal will heat a room faster than a stove built of cast iron or ceramic material. It also means that the sheet-metal stove will not hold the heat as long; after the fire dies, it will cool immediately. On the other hand, a cast-iron stove will hold the heat, though it takes longer to begin passing the heat into the room. It will continue to radiate heat long after the fire has died. Brick, stone and tile have even longer heat-retention qualities. Durability is another factor in the kind of material used to make the stove. Cast iron lasts longer than sheet metal. Another factor is moveability: sheet metal is lighter and therefore easier to move than cast iron or ceramic material.

Cast iron has proved to be the best compromise for most stove needs. It is durable, heavy, but still moveable, and has excellent heat-retention characteristics. Some stoves have combinations of all these materials. I have a large potbelly stove with a cast-iron belly lined with fire bricks. This particular stove has a sheet metal top which radiates immediately. The firebricks hold heat long after the fire is out, and the cast-iron belly continues to radiate warmth long into the night.

Ceramic or stone stoves have never been widely used in this country, though the soapstone stove was popular in Vermont. In Europe, however, large stone stoves with tile facing are common. These structures can hold heat throughout the day. The common practice is to burn a bundle of sticks called a "faggot." The faggot is fashioned out of brush and limbs gathered from the forest and tied together with wire. The sticks close to the center are the smallest. The faggot burns with great intensity for about 45 minutes, then the ashes and glowing embers are raked to the back of the stove and dampers are switched so that the exhaust drafts are routed through many channels in the brick, stone and tile structure. Every bit of heat is absorbed and then slowly radiated into the home throughout the whole day.

COUNTRIES WHERE STOVES ARE PRODUCED

Every type of stove previously discussed is produced all over the world; however, there are certain characteristics common to stoves produced in different parts of the world. My experience keeps me from commenting on all of these stoves, but I have observed the following:

The Scandinavian and Canadian models have excellent workmanship and superb design.

Austrian stoves combine cast iron, fire brick, automatic thermostats, and generally fine quality craftsmanship.

USES

When choosing a stove, the buyer must ask himself, "Where and for what is the stove designed to be used?"

Parlor stoves are designed to heat one room and take up little living space. They are small, compact and efficient. Most parlor stoves are ornately decorated, oblong, and feed from the side. Many have grates so that wood or coal can be used.

The kitchen stove will become the center of the wood-burning home. A well-designed kitchen stove can heat several rooms, cook unlimited pots on its surface, bake anything and everything in its ovens, keep untold goodies warm in its warming ovens, heat water for all hot water needs, dry clothes on its drying racks, dry and smoke edibles for storage, and heat the rest of the house with radiators heated from water in its firebox. Kitchen cook stoves are an absolute must for anyone considering heating by wood.

Finally, in making a stove selection, determine how the stove is to be used as a heater—a full-time heating system, an emergency backup, or a sentimental conversation piece. There are stoves built for all these purposes, and the buyer must match his needs with the designed capacity of the stove.

ACCESSORIES

At one time or another, the following items will be needed:

1) Chimney Sweep. This is available at hardware stores and it comes in a crystalline form. One-half a cup per week thrown on a working fire will keep flue and chimney relatively free of soot and creosote. However, this does not mean that chimney sweep will clear the interior walls of a chimney or flue. It means only that it will act as a preventative.

2) Firescreens. For open stoves and fireplaces, these are a must and should fit tightly.

3) Stove Polish. This is used for maintaining a good stove finish.

4) Rutland Stove Cement. For at-home repairs.

5) A sheet of fireproof material for the stove to rest on. Most hardware stores sell asbestos-lined stove mats, which, aside from being important safety features, are decorative as well.

6) Heat shields for stoves placed near combustible walls. They can be in the form of asbestos mats or corrugated aluminum roofing which will reflect the heat back towards the center of the room, while protecting the wall.

7) A bellows is handy for starting up dying coals.

8) Pokers, shovels and prongs for the dirty work of cleaning out ashes and shifting logs.

9) Fire extinguishers specifically designed for wood fires, which should be purchased and hung within easy reach of all members of the household, before the stove is installed. It is also a good idea to have regular fire drills.

10) A woodbox is nice to have. They come with bottoms or without. The latter allows one to clean more easily come spring by simply lifting the box frame and sweeping. Don't place the

box too close to the stove. Sparks find them to be enticing objects.

11) Bedwarmers—these are stone, brick or ceramic. Put them in the stove oven, then wrap in a towel or heavy material. Slide them under the covers before getting into bed. They will take the chill out of the bed in an unheated bedroom and radiate heat throughout the night. Our family is convinced they are far superior to an electric blanket.

CONNECTING THE STOVE TO THE CHIMNEY

Stove pipe is used to connect the stove to the chimney. It comes in sizes 4 through 10 inches and is fashioned out of light sheet metal. One piece is crimped at one end in order to fit into the larger end of another piece. The crimped ends always point up or away from the stove. The pipes slide together to form tight joints. These joints are further secured using stove cement. Elbows come in a variety of angles so that the pipe can be bent in any direction.

The basic principle for making stove pipe connections is to try to vent the stove as directly as possible into the chimney. The stove must enter the chimney horizontally and should not extend into the flue. The hole in the chimney wall must be lined with fire clay or metal thimbles. The latter are metal collars that fit around the stovepipe and slide up against the wall to make a tight fit at the point where the pipe enters the wall.

Stove pipe should never be closer than 18 inches to woodwork or other combustible material. If less than 18 inches from woodwork, cover the pipe with fire-resistant material. If the pipe must pass through a wooden partition, the woodwork must be protected. Insulated pipes are designed for just this purpose. If none is available, cut a hole 18 inches larger than the diameter of the pipe and then refill it with fire brick.

Stoves should not be connected to flues used for central heating units. Two or more stoves can be connected to the same flue if the flue is large enough and the stove inlets to the flue are more than 3 feet apart. It is much safer to have separate flues for each stove.

Most stoves are designed to operate with dampers installed in the pipe directly over the stove. If the stove has a built-in damper directly in front of the exhaust pipe collar, no damper is needed in the pipe. Dampers are easy to install. They come apart so that the rod is pushed through a predrilled hole, then the disk fits on the rod and the damper is ready to use.

Several other accessories such as backdraft baffles, and draft adjustment valves can be added to the pipe or stove as needed. Usually, good dampers and well-erected stove pipe connections will suffice for the average stove.

USING THE STOVE

The wood stove works on the basis of the combustion that takes place in the firebox. That combustion is fed, increased, retarded or stopped completely by the manipulation of air intake and smoke exhaust valves.

A draft control beneath the firebox or near the floor of the firebox serves to adjust the direct flow of air into the combustion chamber. Controls located on the flue or stovepipe serve to control the hot air pressure expelled from the combustion chamber. These valves differ with each stove depending upon the amount of draft, the air circulation around the stove, and the particular design of the stove.

Types and styles of air intake valves differ widely, but the general rule is that intake valves which let air into the firebox from below or through a grate will produce more efficient combustion.

Another air intake valve which is found on many stoves is a separate valve from the intake combustion valve. This second valve is located above the first and lets air into the firebox above the flame. This valve can be used to either cool the firebox or to assist in producing more efficient combustion. If the valve is opened completely, it will let in large amounts of air which will cool the firebox. The air will go across the top of the flame and up the draft, creating greater draft pressure if the exhaust draft is fully open. The same valve will produce the opposite effect if the fire is burning at a moderate rate with adjusted oxygen coming into the fire from below, and this air valve is set to let in small amounts of air into the combustion chamber. The exhaust valves must also be adjusted to inhibit the exhaust draft pressure. The correct adjustment of this valve will assist in a secondary burning of gases that were not consumed in the primary burning. Such adjustment will decrease the pollution caused by wood burning, and will produce greater heat. The ability to produce this secondary burning means learning to balance the valves on each stove and flue.

The following steps are the process of starting a fire in a stove.

1) Clean all ashes from the floor or grate of the firebox.

2) Make sure that all exhaust and intake valves are open.

3) Crumple newsprint or other paper which burns easily on the grate or the floor of the firebox.

4) Cover this paper with dry kindling. The kindling should not be packed tightly, but should be allowed to rest loosely on the paper. There should be air space around each piece of kindling. The kindling should graduate in size with the largest being at the top of the firebox and the smallest being next to the paper.

5) Light the paper in several places and watch the burning to insure that the kindling catches on fire. Once the kindling is burning vigorously, begin to shut down the vents. Watch the fire to determine the effect of valve adjustment on the combustion.

Fires which are burning too rapidly will smoke. Smoke produced too rapidly to be pulled out the draft will puff out into the room. To correct this situation, close the bottom intake valve until the smoke clears and the chimney vents correctly. Then gradually adjust the valve to an efficient open position. If the fire is burning too slowly, this can be corrected by letting in more air from below or by opening the draft on the flue to create more pressure which will pull in more air to the firebox.

NEVER use any sort of flammable liquid to assist the fire. Stoves are not designed for quick starts and such misuse will damage the firebox and may cause explosions.

When the fire is burning steadily, close the doors and adjust the draft until the fire burns evenly.

Cooking on Wood Stoves

Once fire control is learned, the stove owner is ready to try cooking. Even the most modest heating stove is usually equipped with a cooking pot lid on the top, and so gives an opportunity for some experimenting in the art of wood stove cooking. A good cook learns how to create hot fires, slow fires or moderate fires, and how to use each type in cooking.

Good wood stove cooking depends on good cookware, and the best is cast iron cookware. Most foundries which make wood stoves also make cast iron cookware as a sideline, and a great variety is available. There are also several nationally known brands.

The most basic item of cookware is the tea kettle. The large cast iron kettle which should be used on a wood stove is not very similar to its modern copper-bottomed aluminum cousin which whistles shrilly only when the kettle boils. Cast iron kettles have crooked or "serpentine" spouts, and when set to the back of the stove will gently sing all day long with a light laughing sound. The kettle is always just ready to boil and when moved directly over the fire, will immediately do so. It should also be mentioned that keeping a kettle on the back of your stove, whether or not it is a cook stove, will provide added moisture in the room which increases the efficiency of the heat and serves the function of the modern humidifier at no added cost.

Stoves which are specifically designed as kitchen ranges offer far wider possibilities than the heating stove with one pot hole in the top, but even these simple varieties can be used to keep a one-pot dish such as soup or stew simmering all day, while the kettle is using up the back of the stove. A coffee pot will also stay warm without reboiling the coffee if left on the one-holer. The basic idea is to learn how to best utilize energy. If a fire is burning, learn to put on soup for simmering and thereby utilize energy which otherwise would be used only for heating.

The same concept of energy utilization also applies to the use of the more versatile kitchen range. Warming or simmering pots can be put furthest from the flame and quick boiling can be accomplished at the same time directly over the flame. Most ranges come with draft controls which cause the flame and resulting heat to be directed to different places on the top of the stove as needed. Surface cooking is a process of learning to control the heat, and also of learning how to gauge the heat quality at different places on the stove.

For boiling, the stove has several important features. A well-designed range has cooking holes which are located directly over and near the fire. At least one of these holes should have a cover consisting of concentric circles which can be removed individually. Skillful use of this arrangement provides for heat adjustment. For most rapid boiling, the pot should fit right into the hole, taking the place of the cover, since the more of the pot that is exposed directly to the heat, the quicker the boiling will occur. If one or more of the smaller circles are left in place, the bottom of the pot is less exposed to flame and will heat more slowly. A pot used for boiling should always be covered for most rapid boiling.

The surface of the stove can also be used for broiling or making toast. Toast made on the surface of a wood stove is the best there is. A slice of bread (of course homemade bread is best) is put on the stove top directly over the fire. When cooked on one side, it is flipped with a knife and the other side is cooked. No modern toaster can come close to producing this crisp delicacy.

Broiling can also be done directly on the surface, and many ranges have a special side vent which swings down to allow for insertion of a rack directly over the coals. The fire must be properly prepared

  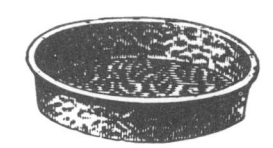

for broiling. Coals should reach a uniform color of irridescent red, which indicates that most of the gaseous combustion has occurred and that the coals are all at the same temperature. The draft should be adjusted to let plenty of oxygen enter the firebox. If a dark or dead spot occurs in the coals, the meat will not be cooked over this spot. Some cooks prefer not to use a rack for broiling. They prepare a fire of coals made from the same kind of hardwood, and then break up all the coals until there is a smooth even bed of coals all the same color. The meat is placed directly on the coals to sear the meat on both sides. Frequent turning is required to prevent burning. The searing prevents escape of juices which flavor the meat. Olive oil lightly rubbed over the meat helps prevent burning and adds a delicious flavor to the broiled meat. The rack over the fire will slow the process but will also offer greater protection from burning. One advantage of cooking directly on the coals or on the rack is that it saves cleaning the surface of the stove.

The real test of a wood range and of the cook's skill is baking. Learning how to control the oven heat in a wood stove is a matter of trial and error. Most ovens are equipped with thermometers which help in the learning process, but real skill must be acquired in the manipulation of the draft controls. When the fire first starts, the flames should rush over the top of the oven and heat this area. The range is equipped with a damper that will stop this escape route and force the flames down and under the oven to heat the bottom. Skillful use of the damper can result in an even heat.

Baking should be done on the bottom floor and not on the rack, since the middle of the oven will generally be too cool for good baking. Each cook must discover what works best for his or her particular stove, but it is basically a matter of learning fire control and knowing the functions of the drafts designed into the stove.

Many modern cooks will be frightened by the absence of an accurate thermometer or thermostat on the oven since they are used to recipes that specify a temperature for a particular item to be baked, and an oven that automatically provides this temperature. There is no need for this fear, since a little common sense will solve both problems. The original Fanny Farmer cookbook, the ancestor of one of today's most popular works in the field, didn't specify degree settings for the baking recipies. It only specified "hot," "medium" or "cool" temperatures for the oven. The cook can make that decision today in relation to the degree setting given in the recipe. The cook also acts as the thermostat, learning to rotate the item being baked and to control the stove drafts around the oven to maintain the desired heat in it. Most experienced cooks claim that anything baked in a wood stove oven is better tasting because the flavor is not vented out of the oven. Only wood stove ovens are unvented.

The wood stove has other advantages which just do not exist in the modern electric or gas model. Most stoves have a warming shelf or shelves at the back of the range which provide a perfect place for raising the dough for baking bread (for that toast) or any other yeast recipe. The shelf is perfectly placed so that the dough will get the necessary amount of heat but not too much. These shelves have a multitude of other uses. They will keep a whole meal warm if necessary. They provide a perfect place for drying mittens, gloves, socks, and anything else needed in a hurry. People who get used to drying wet clothes over a wood range begin to wonder how other people get through the winter. The shelves are also handy for storing food and condiments which need to be kept dry, such as salt.

Finally, the shelves are the perfect winter napping place for a cat and there is usually one sleeping here. For all I know, they may come from the factory that way.

Cooking ranges can be complicated or simple. Some are cabinet style and some are baroque. Some have accessories and gadgets which are extremely practical, such as hot water heating coils running around the firebox and hot water reservoirs. Some have glass doors for the ovens, and one stove I saw recently had an air-tight ash pit which emptied through the floor into a specially designed receptacle in the cellar. A number of stoves have a space directly over the oven and under the top surface which can be used to dry orange and lemon peels for making kindling. Their natural oils catch quickly and burn for a long period.

A wood cooking range must be cleaned once a month if it is being used regularly. Soot collects in all the draft passages and must be removed in order to maintain peak efficiency. This is a messy job. The most difficult place to clean is under the oven. There is usually a cleaning hole through which a long bent wire is pushed into all the crannies and cracks to loosen the soot. All of this soot must then be pushed back to the hole and cleaned out.

Another important accessory of the kitchen range of earlier days was the "sad" iron, used for ironing clothes. It was heated in the stove. A set of three Sadirons consisted of one, two, and three pound irons which had interchangable handles and which were detached for heating in some cases. The irons were heavy cast iron and were put directly over the firebox to absorb the maximum amount of heat before being used. Two could be heating while the third was being used. The heavier irons were used for heavy fabrics while the one-pounders were designed to iron lighter materials such as fine cotton.

How to Build Practical Wood Stoves

The most commonly used home-built stove is made out of a 55-gallon oil drum. Such a stove can be very simply fashioned by cutting a hole in the front for feeding the fire and a hole in the top for venting the smoke. The drum can be rested on its side or positioned upright (See Figure 1). It can stand on legs or be hung from the ceiling by chains.

The more complex type of 55-gallon drum stove can be fashioned by adding already manufactured parts, such as a cast iron door with an automatic draft control mounted on it. The drum can be lined with fire bricks and complicated grate systems can be added to allow for the recycling of excess gases and to preheat the air entering the stove. It is also relatively easy to make a heat exchanger out of another 55-gallon drum by placing it over the original drum. (See Figure 2)

THE DOOR

This can be a simple hole or a precast factory item. A number of companies make doors aimed specifically at the 55-gallon drum stove. Many can be mounted directly or modified to fit the drum.

To mount a flat door, it is easier to use the drum stove resting on its side. The door can then be mounted on either one of the flat bottoms.

If the stove is to stand upright and a flat door is to be used, build a layer of metal strippings around the frame of the cut-out until the flat door fits tightly. (See Figure 3)

A simpler method is to use the piece of drum that comes from the cut-out. Enlarging it by adding metal strips along its outside edges and remounting it with hinges on one side will make a good door. (See Figure 4)

Placement of the door is critical. If the door which is to be mounted has an air intake draft control, then it should be positioned as close as is possible to the grate level.

If wood is the intended fuel, and no grate is to be used, mount the door 2 to 3 inches from the bottom of the drum. Sand and gravel can be placed in the bottom of the drum to extend its lifespan. This also increases its heat holding capacity. If a free standing cast iron grate is to be used, the door can be positioned slightly higher than the 2 or 3 inches suggested earlier. Be sure, however, that air can enter easily under the grate to feed the fire. Simple grates made from reinforcement rods can be used. Simply stick the rods through the sides of the drum about a foot off the bottom. In this case, the door can be positioned 2 inches above the grate level.

The drum stove can be further improved by purchasing a cast iron door with a thermostat mounted on it. This device will make the drum an airtight automatic stove.

Figure 1. 55-gallon drum stove on side and on end.

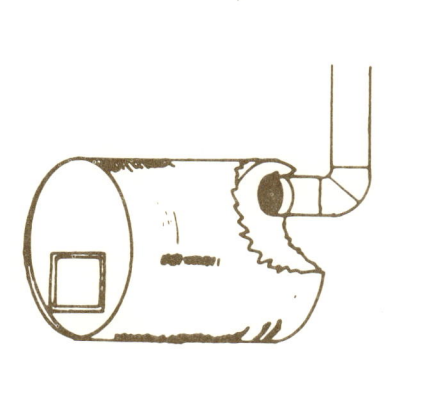 

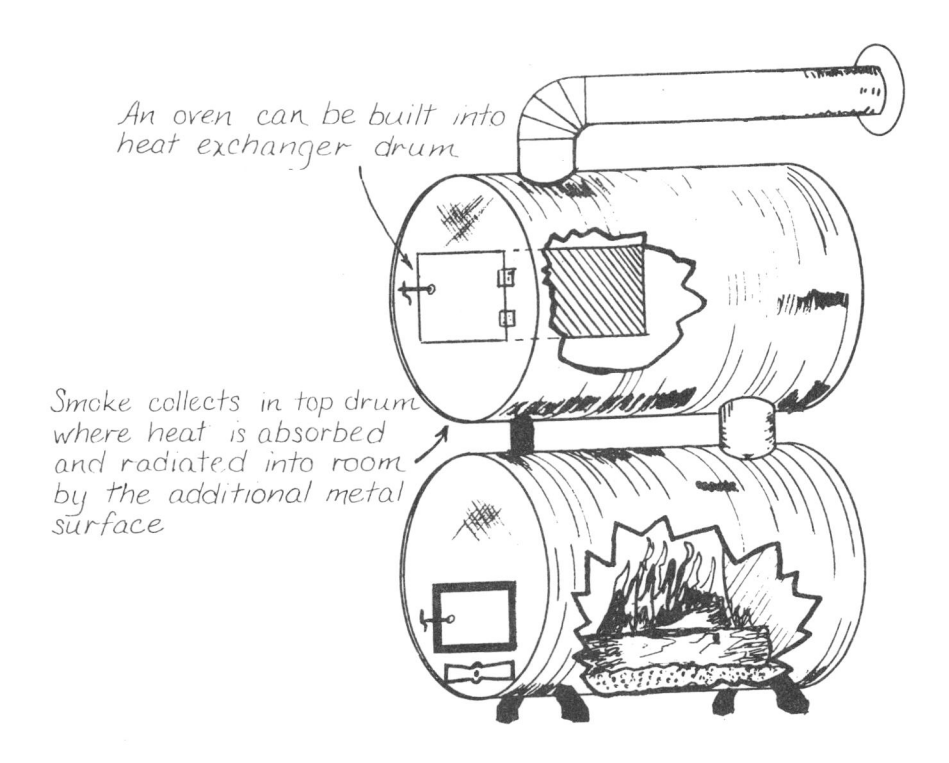

An oven can be built into
heat exchanger drum

Smoke collects in top drum
where heat is absorbed
and radiated into room
by the additional metal
surface

Figure 2. *55-gallon drum smoke exchanger with built-in oven.*

Heavy gauge
metal, welded
to drum

Air intake vent

Figure 3. *A flat door mounted on the side of drum, using heavy-gauge metal reinforcements.*

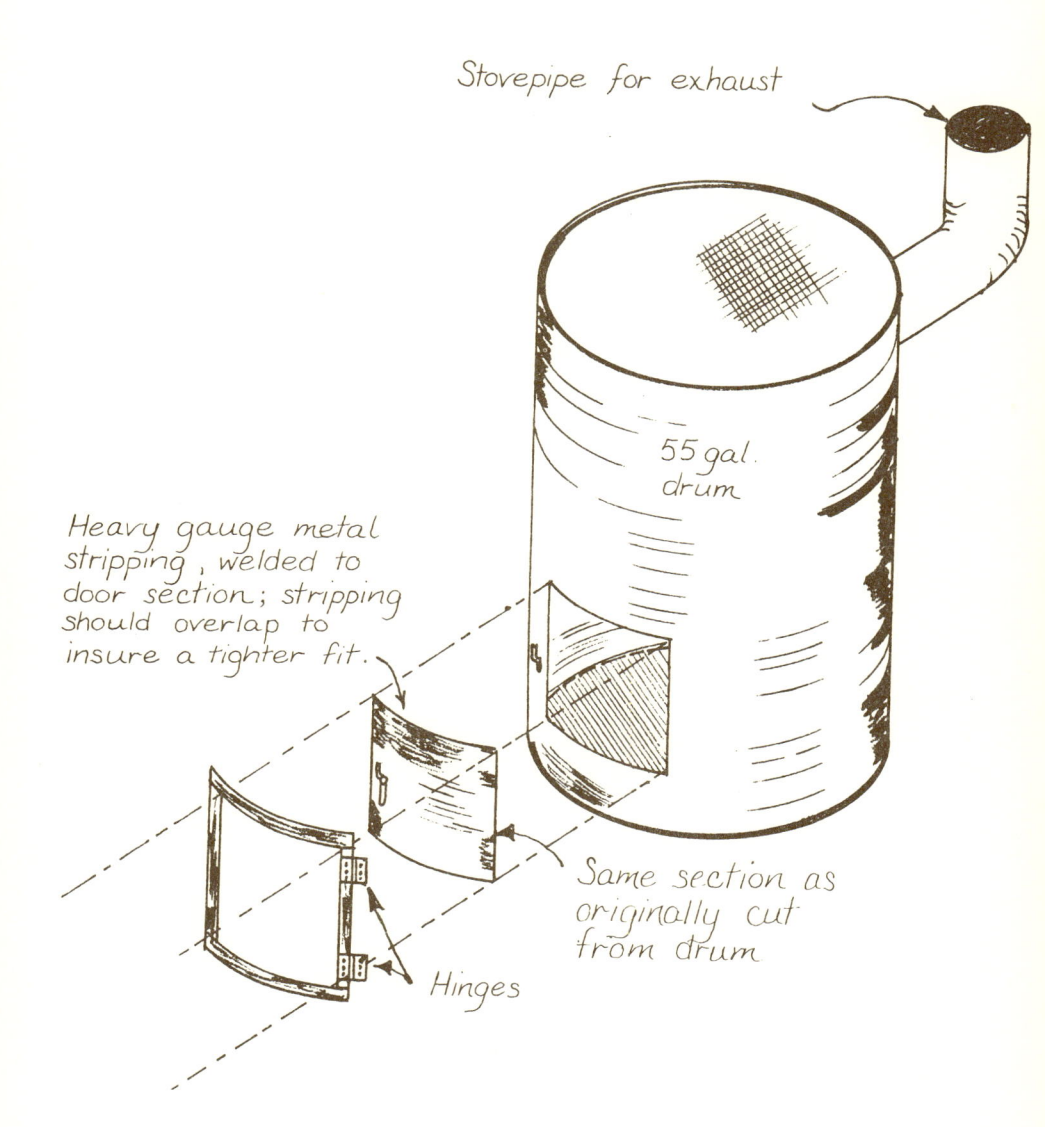

Stovepipe for exhaust

55 gal. drum

Heavy gauge metal stripping, welded to door section; stripping should overlap to insure a tighter fit.

Same section as originally cut from drum

Hinges

Figure 4. Standing 55-gallon drum stove with door made from cutout section of drum.

STOVE LINING

To improve heating efficiency, line the drum with bricks. (See Figure 5) This process means removing one end of the drum so that the bricking process can be accomplished. Some drums have removable lids and it is best to stick with them, as shearing a welded end leads to problems when the cutout is put back after the lining process is completed.

The lining should be of fire brick, stove liner (available at hardware stores), or furnace clay. The lining need only extend 18 inches above the grate level. Each layer of fire brick added improves heating efficiency.

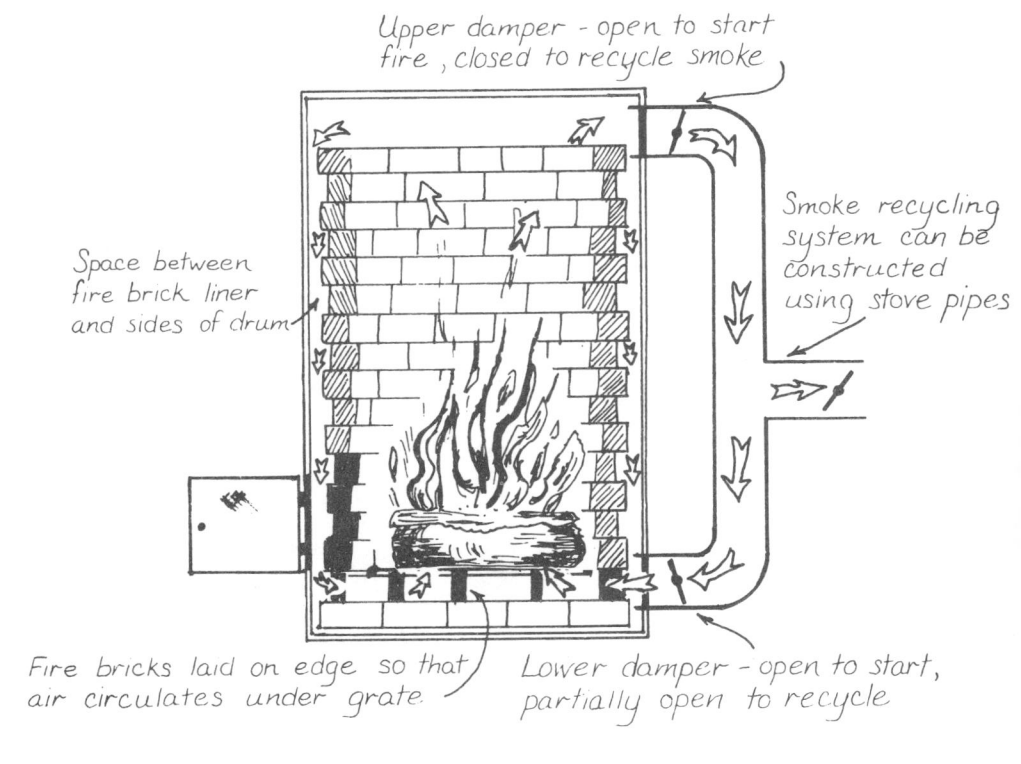

Upper damper - open to start fire, closed to recycle smoke

Smoke recycling system can be constructed using stove pipes

Space between fire brick liner and sides of drum

Fire bricks laid on edge so that air circulates under grate.

Lower damper - open to start, partially open to recycle

Figure 5. Standing 55-gallon drum stove with fire brick liner and recycling system made from stove pipe. Note the dampers mounted at the top and bottom of the stove pipe and vent system under grate.

EXHAUST VENT

The hole for the smoke draft should be mounted opposite the air intake at the back or on the top. If the stove rests upright, the exhaust vent should be placed on the opposite side from the door opening and at the top. If the stove is designed to burn wood without a door, the pipe must be at least 8 inches in diameter.

The pipe can be attached by simply making a hole, inserting the stovepipe and filling in the seams with stove cement in order to make a tight fit. A more sophisticated connection can be made by welding a collar around the opening which has an outside diameter that matches the inside diameter of the pipe itself. Thus, the stove pipe fits over this collar to create a perfect seal.

GRATES

Grates for the firebox can be fashioned out of reinforcement rods. These can be bent to the shape of the drum or stuck through the sides of the drum to form a platform. Angle iron or almost any type of sheet metal can be used. In stoves that are to be lined the grates will have to be strong enough to hold the additional weight of fire bricks. Grates used for drum stoves with recycling features will need the additional inclusion of air vents behind the brick liners.

HEAT EXCHANGER

A heat exchanger can be added to make use of the heat normally vented up the flue. Another drum can be used. (See Figure 2) The stove pipe is run from the stove drum into the heat exchanger drum.

The smoke collects in the latter drum and is then vented into the chimney. This extra drum absorbs much of the heat contained in the smoke and adds greatly to the surface area radiating heat into the room.

RECYCLING SYSTEMS

The recycling system improves the combustion efficiency of a stove by burning gases given off during the initial burning of wood. This is accomplished by shutting the exhaust draft (that feeds smoke to the chimney) and forcing the gases down into the fire again. The draft in the bottom of the stove is connected to passages which lead from the firebox chamber to an area under the grate. The fire pulls air from this chamber for combustion, thus creating a partial vacuum. This creates a difference in pressures between that which builds up in the firebox and under the firebox. This pressure will pull the exhaust gases down under the grate where they will mix with incoming air and be drawn a second time into the fire. (See Figure 5)

This type of system can be built into a drum stove that utilizes a grate system and a lining. The brick lining should extend 18 inches up from the grate. A 1-inch space is needed between the lining and the side of the drum itself. This air space must be vented through an area below the grate. Small holes drilled all around the floor of the firebox between the brick lining and the drum itself will suffice for such venting. The air intake will also need to vent directly into this area below the grate. A round hole for the exhaust draft will also lead from this area. Install the stove pipe in this hole and mount a damper in the pipe. The system is now complete.

To work it, once the fire is burning steadily, shut both exhaust dampers, the one at the top and the one below the grate. The gases will start their recycling process. Eventually, smoke will start coming out of the air intake draft. Open the lower exhaust draft damper. The correct adjustment will produce a steady burning and recycling of the gases. Excess gases will vent into the chimney.

AIR INTAKE VENTS

These vents are for adjusting the amount of air entering the firebox. They may be in the door or under the grate level. The two easiest kinds of vents are sliding and revolving. A sliding vent is made by cutting a horizontal slit through which a bolt will fit which has a large washer on the inside. The bolt holds the sliding vent plate which will have holes the same size and corresponding to the holes cut in the side of the drum. These holes must be far enough apart to insure that all holes will be covered when the slide is at one side and totally open when the slide is at the opposite side. The total area of the holes need not exceed 10 square inches. Guides for the sliding plate can be made for each side of the plate by bolting small bolts to the drum so that the bottom of the slide rests on the bolts.

A revolving vent is made in a similar manner except that the outside plate revolves to open and close the vents. The vent plate revolves on a bolt in the center of the air vent holes. Again, be sure that the plate revolves so that the holes can be totally opened and totally closed.

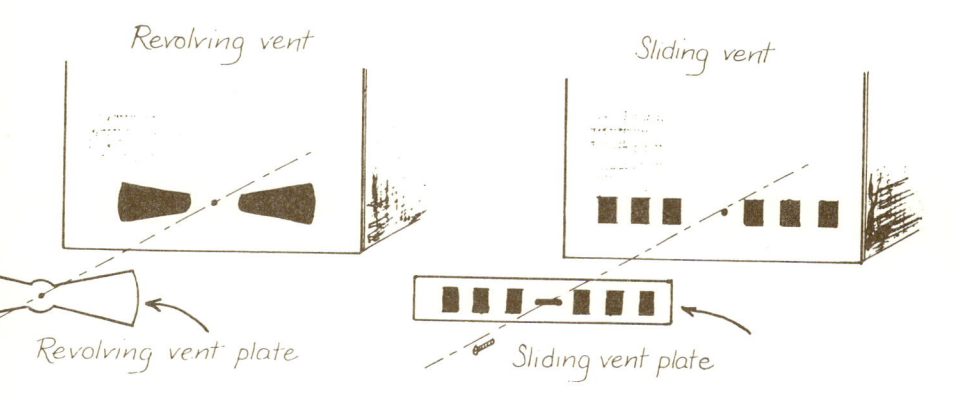

Figure 6. Revolving and sliding air intake vents.

THE BOX STOVE

The box stove is made from six rectangular plates of sheet metal which are bolted together with angle irons to form a box with a door at one end. The top, bottom and sides are bolted together with the angle irons on the inside of the stove. The next step is to bolt the back piece in place.

The legs are bolted to the outside and form the vertical joints of the box. The door is made from sheet metal and is hinged to the legs.

When the entire box is bolted together, seal the seams with stove cement.

The air intake is then mounted on the door. Cut small holes for the intake opening and then cut a piece of sheet metal for the outside hole covering. This can be made to either slide back and forth across the hole to allow air in or to shut it out, or it can be circular.

The exhaust vent is cut through the top place at the rear of the stove. This joint can also be made tight by using stove cement.

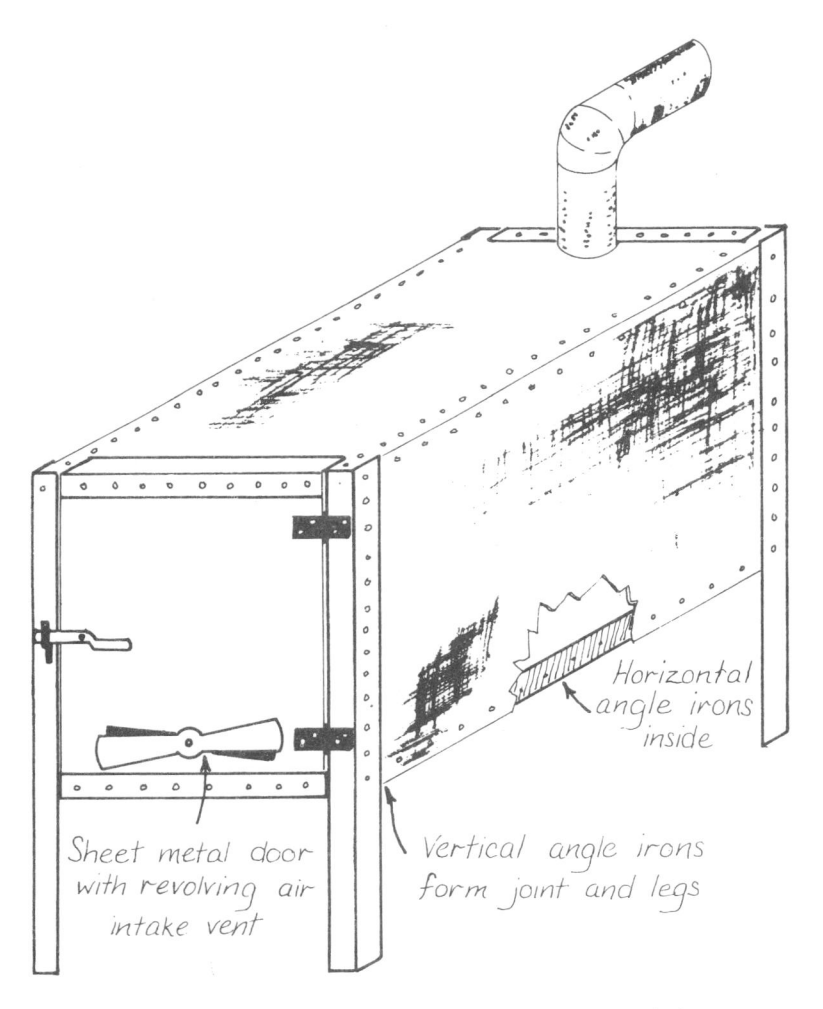

Figure 7. A box stove built from sheet metal and angle irons.

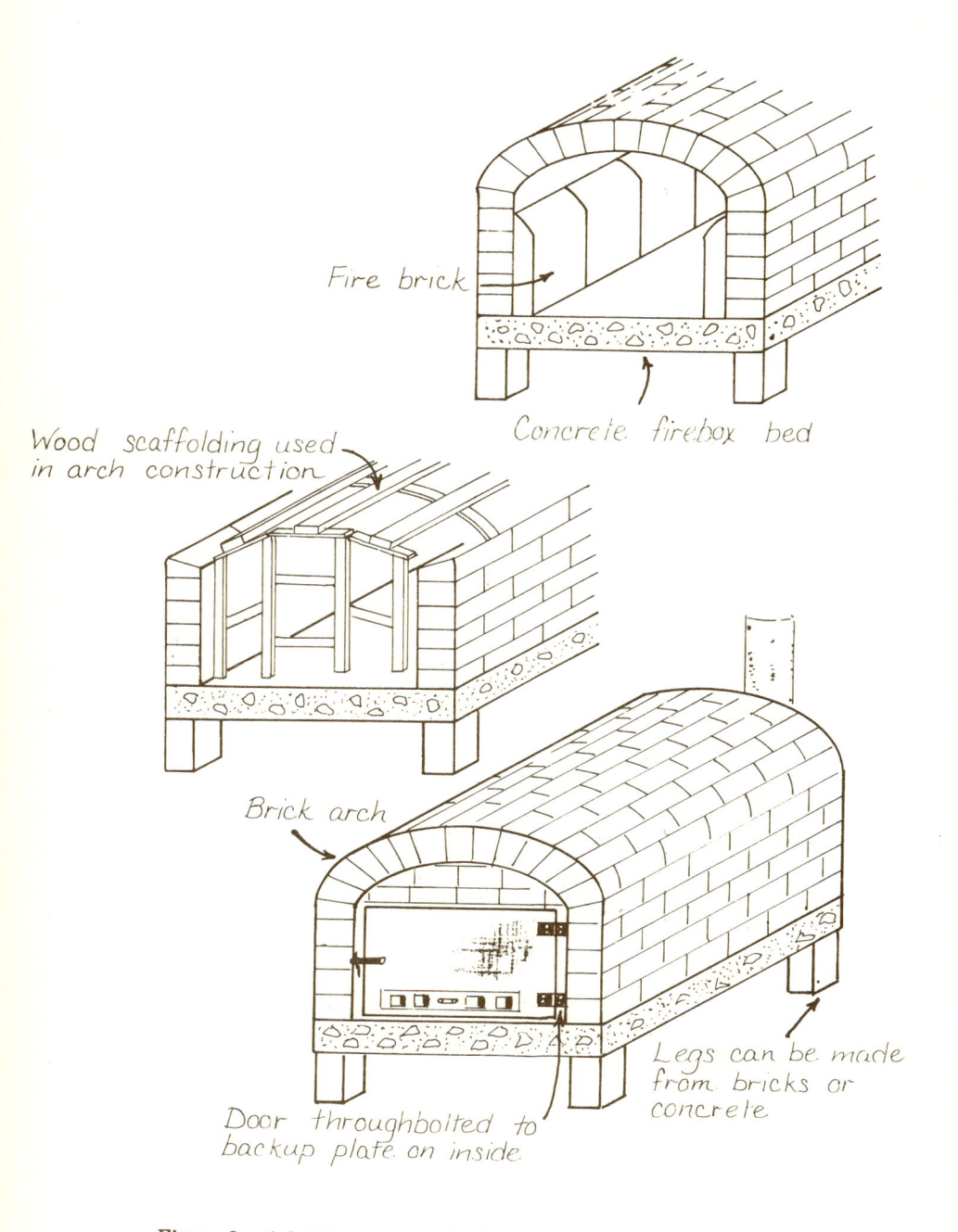

Fire brick

Concrete firebox bed

Wood scaffolding used in arch construction

Brick arch

Door throughbolted to backup plate on inside

Legs can be made from bricks or concrete

Figure 8.   A brick arch stove built on a concrete base.

THE BRICK STOVE

A brick stove can be constructed by using cement for the foundation and the legs. A wooden form for the legs and foundation is made on the spot where the stove will rest, as it is a permanent structure. Reinforcement rods are used inside the concrete and should extend into the legs to insure that they are well secured to the structure. The forms are not removed until the concrete is "cured" or fully dried and settled. When the foundation and the legs are ready, the sides of the stove are built and the interior is lined with fire brick. Fire clay is used to secure the fire brick.

Building the arch is a difficult task and can be made easier by first erecting an internal scaffolding that predetermines the exact shape desired.

The end walls fit under the sides and will need cutting in order to achieve a sound fit. A special broad chisel makes this job easier. The front wall will need a square opening for the door. The bricks above this door can be supported by an angle iron across the entire wall.

The hinges for the door can be bolted directly into the brick wall if a metal backup plate is used on both the interior and the exterior walls. Fire clay, mixed to a putty consistency is then applied around the door edges. Shut the door while the clay is still soft and this will make an impression of the door edges which will harden and allow for a tight fit when the door is closed on the fully constructed stove.

The smoke hole at the top rear can be chiseled out once the arch is completed or a hole may be left during construction. Fire clay can be used to create a seamless connection once the stove pipe leading to the chimney has been inserted.

Further elaboration on the process of building such a stove would prove futile due to the many masonry techniques involved. It is suggested that if one finds a brick stove to be enticing, thorough reading should be done on masonry techniques. It is a process that demands concentration and a gradual development of skills. Too, we would hate to encourage construction of a structure that may at a later date prove to be hazardous due to gaps, weaknesses and so forth. This type of stove is highly effective and efficient. Think of it as a major undertaking and the process will prove to be rewarding, resulting in a safe product.

Chimneys New and Used

All stoves, fireplaces and other wood-burning equipment require some type of chimney (Figure 1). The chimney must be designed and built so that it produces sufficient draft to supply an adequate quantity of fresh air to the fire and to expel smoke and gases emitted by the fire or equipment.

A chimney located entirely inside a building has better draft than an exterior chimney, because the masonry retains heat longer when protected from cold outside air.

FLUE SIZE

The flue is the passage in the chimney through which the air, gases and smoke travel.

Proper construction of the flue is important. Its size (area), height, shape, tightness, and smoothness determine the effectiveness of the chimney in producing adequate draft and in expelling smoke and gases. Soundness of the flue walls may determine the safety of the building should a fire occur in the chimney.

Manufacturers of fuel-burning equipment usually specify chimney requirements, including flue dimensions, for their equipment. Follow their recommendations.

HEIGHT

A chimney should extend at least 3 feet above flat roofs and at least 2 feet above a roof ridge or raised part of a roof within 10 feet of the chimney. A hood (Figure 2,C) should be provided if a chimney cannot be built high enough above a ridge to prevent trouble from eddies caused by wind being deflected from the roof. The open ends of the hood should be parallel to the ridge.

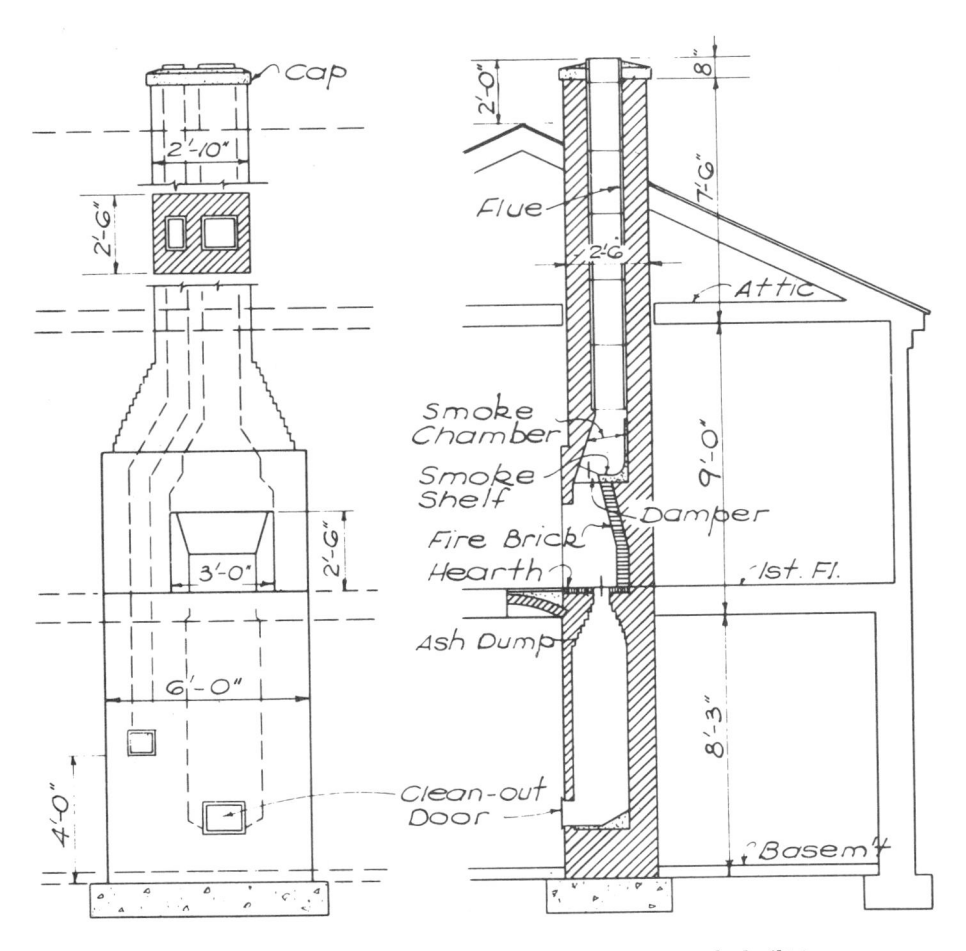

Figure 1. Diagram of an entire chimney such as is commonly built to serve the house-heating unit and one fireplace.

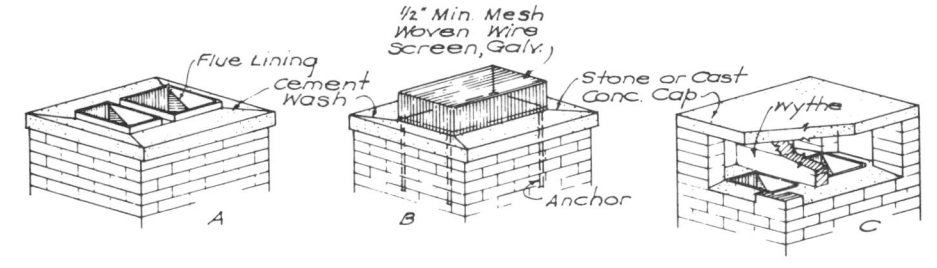

Figure 2. Top construction of chimneys. A, Good method of finishing top of chimney; flue lining extends 4 inches above cap. B, Spark arrester or bird screen. C, Hood to keep out rain.

Low-cost metal-pipe extensions are sometimes used to increase flue height, but they are not as durable or as attractive as terra cotta chimney pots or extensions. Metal extensions must be securely anchored against the wind and must have the same cross-sectional area as the flue. They are available with a metal cowl or top that turns with the wind to prevent air from blowing down the flue.

SUPPORT

The chimney is usually the heaviest part of a building and it must rest on a solid foundation to prevent differential settlement in the building.

Concrete footings are recommended. They must be designed to distribute the load over an area wide enough to avoid exceeding the safe load-bearing capacity of the soil. They should extend at least 6 inches beyond the chimney on all sides and should be 8 inches thick for one-story houses and 12 inches thick for two-story house having basements.

If there is no basement, pour the footings for an exterior chimney on solid ground below the frostline.

If the house wall is of solid masonry at least 12 inches thick, the chimney can be built integrally with the wall and, instead of being carried down to the ground, it can be offset from the wall enough to provide the flue space by corbelling. The offset should not extend more than 6 inches from the face of the wall--each course projecting not more than 1 inch--and should be not less than 12 inches high.

Chimneys in frame buildings should be built from the ground up, or they can rest on the building foundation or basement walls if the walls are of solid masonry 12 inches thick and have adequate footings.

Local codes may call for slightly different construction of fireplaces and chimneys than is given in this article. In such cases, local code requirements should be followed.

FLUE LINING

Chimneys are sometimes built without flue lining to reduce cost, but those with lined flues are safer and more efficient.

Lined flues are definitely recommended for brick chimneys. When the flue is not lined, mortar and bricks directly exposed to the action of flue gases disintegrate. This disintegration plus that caused by temperature changes can open cracks in the masonry, which will reduce the draft and increase the fire hazard.

Flue lining must withstand rapid fluctuations in temperature and the action of gases. Therefore, it should be made of vitrified fire clay at least five-eighths of an inch thick.

Both rectangular- and round-shaped linings are available. Rectangular lining is better adapted to brick construction, but round lining is more efficient.

Each length of lining should be placed in position--set in cement mortar with the joint struck smooth on the inside--and then the brick laid around it. If the lining is slipped down after several courses of brick have been laid, the joints cannot be filled and leakage will occur. In masonry chimneys with walls less than 8 inches thick, there should be space between the lining and the chimney walls. This space should not be filled with mortar. Use only enough mortar to make good joints and to hold the lining in position.

Unless it rests on solid masonry at the bottom of the flue, the lower section of lining must be supported on at least three sides by brick courses projecting to the inside surface of the lining. This lining should extend to a point at least 8 inches under the smoke pipe thimble or flue ring (see Figure 6).

Flues should be as nearly vertical as possible. If a change in direction is necessary, the angle should never exceed 45 degrees (Figure 3). An angle of 30 degrees or less is better because sharp turns set up eddies which affect the motion of smoke and gases. Where a flue does change directions, the lining joints should be made tight by mitering or cutting equally the ends of the adjoining sections. Cut the lining before it is built into the chimney; if cut after, it may break and fall out of place. To cut the lining, stuff a sack of damp sand into it and then tap a sharp chisel with a light hammer along the desired line of cut.

When laying lining and brick, draw a tight-fitting bag of straw up the flue as the work progresses to catch the material that might fall and block the flue.

WALLS

Walls of chimneys with lined flues and not more than 30 feet high should be at least 4 inches thick if made of brick or reinforced concrete and at least 12 inches thick if made of stone.

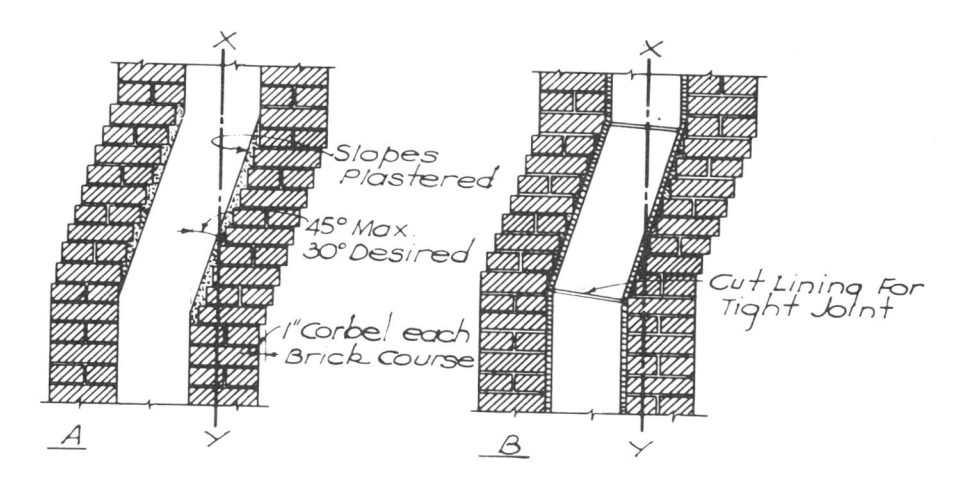

Figure 3. Offset in a chimney. For structural safety the amount of offset must be limited so that the center line, XY, of the upper flue will not fall beyond the center of the wall of the lower flue. A. Start the offset of the left wall of an unlined flue two brick courses higher than the right wall so that the area of the sloping section will not be reduced after plastering. B. Method of cutting lining to make a tight joint.

Flue lining is recommended, especially for brick chimneys, but it can be omitted if the chimney walls are made of reinforced concrete at least 6 inches thick or of unreinforced concrete or brick at least 8 inches thick.

A minimum thickness of 8 inches is recommended for the outside wall of a chimney exposed to the weather.

Brick chimneys that extend up through the roof may sway enough in heavy winds to open up mortar joints at the roof line. Openings to the flue at that point are dangerous, because sparks from the flue may start fires in the woodwork or roofing. A good practice is to make the upper walls 8 inches thick by starting to offset the bricks at least 6 inches below the underside of roof joists or rafters (Figure 4).

Chimneys may contain more than one flue. Building codes generally require a separate flue for each fireplace, furnace, or boiler. If a chimney contains three or more lined flues, each group of two flues must be separated from the other single flue or group of two flues by brick divisions or wythes at least 3¾ inches thick (Figure 5). Two flues grouped together without a dividing wall should have the lining joints staggered at least 7 inches and the joints must be completely filled with mortar.

If a chimney contains two or more unlined flues, the flues must be separated by a well-bonded wythe at least 8 inches thick.

SOOT POCKET AND CLEANOUT

A soot pocket and cleanout (Figure 6) are recommended for each flue.

Deep soot pockets permit the accumulation of an excessive amount of soot, which may take fire. Therefore, the pocket should be only deep enough to permit installation of a cleanout door below the smoke pipe connection. Fill the lower part of the chimney--from the bottom of the soot pocket to the base of the chimney--with solid masonry.

The cleanout door should be made of cast iron and should fit snugly and be kept tightly closed to keep air out.

A cleanout should serve only one flue. If two or more flues are connected to the same cleanout, air drawn from one to another will affect the draft in all.

SMOKE PIPE CONNECTION

No range, stove, fireplace, or other equipment should be connected to the flue for the central heating unit. In fact, as previously indicated, each unit should be connected to a separate flue, because if there are two or more connections to the same flue, fires may occur from sparks passing into one flue opening and out through another.

Smoke pipes from furnaces, stoves, or other equipment must be correctly installed and connected to the chimney for safe operation.

A smoke pipe should enter the chimney horizontally and should not extend into the flue (Figure 6). The hole in the chimney wall should be lined with fire clay, or metal thimbles should be tightly built into the masonry. (Metal thimbles or flue rings are available in diameters of 6, 7, 8, 10, and 12 inches and in lengths of 4½, 6, 9, and 12 inches.) To make an airtight connection where the pipe enters the wall, install a closely fitting collar and apply boiler putty, good cement mortar or stiff clay.

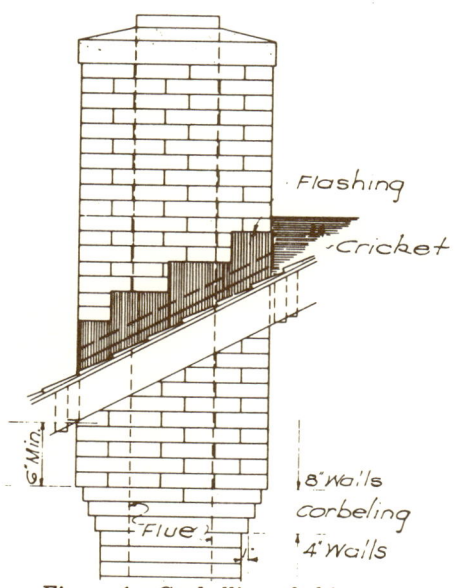

Figure 4. Corbelling of chimney to provide 8-inch walls for the section exposed to the weather.

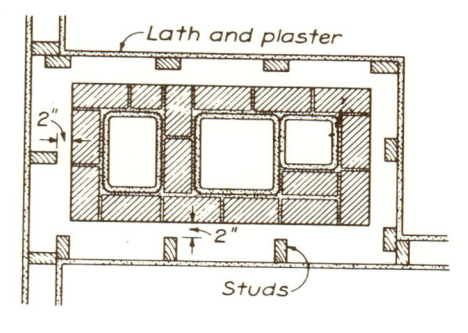

Figure 5. Plan of chimney showing proper arrangement of three flues. Bond division wall with sidewalls by staggering the joints of successive courses. Wood framing should be at least 2 inches from brickwork.

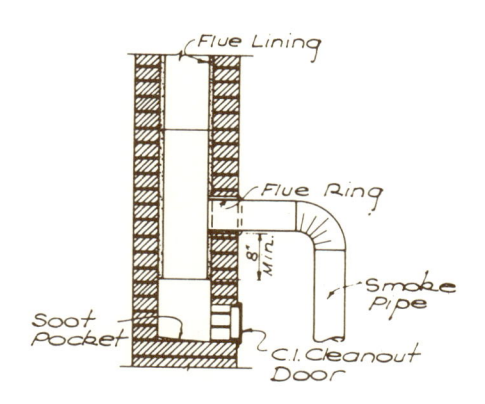

Figure 6. Soot pocket and cleanout for a chimney flue.

A smoke pipe should never be closer than 9 inches to woodwork or other combustible material. If it is less than 18 inches from woodwork or other combustible material, cover at least the half of the pipe nearest the woodwork with fire-resistant material. Commercial fireproof pipe covering is available.

If a smoke pipe must pass through a wood partition, the woodwork must be protected. Either cut an opening in the partition and insert a galvanized-iron, double-wall ventilating shield at least 12 inches larger than the pipe (Figure 7) or install at least 4 inches of brickwork or other incombustible material around the pipe.

Smoke pipes should never pass through floors, closets or concealed spaces or enter the chimney in the attic.

Each summer when they are not in use, smoke pipes should be taken down, cleaned, wrapped in paper, and stored in a dry place.

When not in use, smoke-pipe holes should be closed with tight-fitting metal flue stops. Do not use papered tin. If a pipe hole is to be abandoned, fill it with bricks laid in good mortar. Such stopping can be readily removed if necessary.

INSULATION

No wood should be in contact with the chimney. Leave a 2-inch space (Figure 5) between the chimney walls and all wooden beams or joists (unless the walls are of solid masonry 8 inches thick, in which case the framing can be within one-half inch of the chimney masonry).

Fill the space between wall and floor framing with porous, non-metallic, incombustible material, such as loose cinders (Figure 8). Do not use brickwork, mortar, or concrete. Place the filling before the floor is laid, because it not only forms a firestop but also prevents the accumulation of shavings or other combustible material.

Flooring and subflooring can be laid within three-fourths inch of the masonry.

Wood studding, furring, or lathing should be set back at least 2 inches from chimney walls. (Plaster can be applied directly to the masonry or to metal lath laid over the masonry, but this is not recommended because settlement of the chimney may crack the plaster.) A coat of cement plaster should be applied to chimney walls that will be encased by wood partition or other combustible construction.

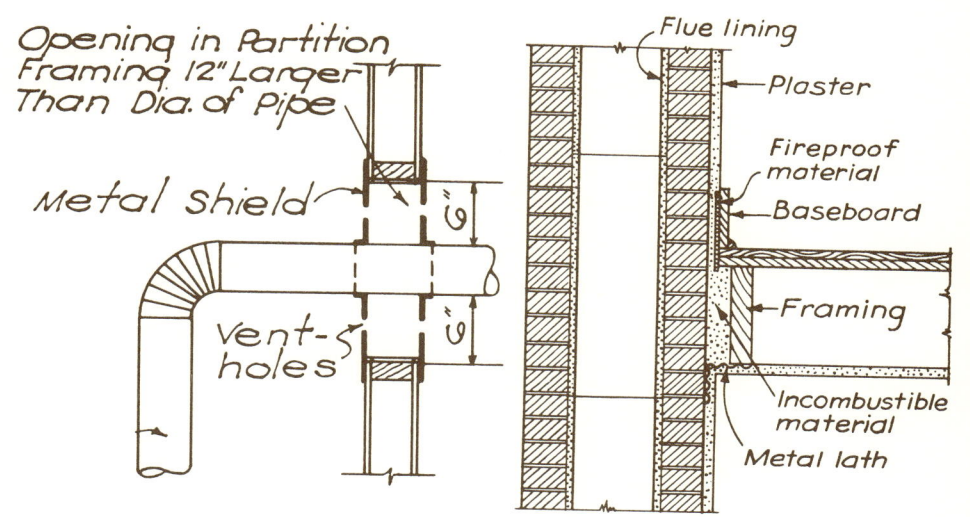

Figure 7. One method of protecting a wood partition when a smoke pipe passes through it.

Figure 8. Method of insulating wood floor joists and baseboard at a chimney.

If baseboards are fastened to plaster that is in direct contact with the chimney wall, install a layer of fireproof material, such as asbestos, at least one-eighth inch thick between the baseboard and the plaster (Figure 8).

CONNECTION WITH ROOF

Where the chimney passes through the roof, provide a 2-inch clearance between the wood framing and the masonry for fire protection and to permit expansion due to temperature changes, settlement, and slight movement during heavy winds.

Chimneys must be flashed and counterflashed to make the junction with the roof watertight (Figures 9 and 10). When the chimney is located on the slope of a roof, a cricket (Figure 9,J) is built as shown in Figure 11 high enough to shed water around the chimney. Corrosion-resistant metal, such as copper, zinc, or lead, should be used for flashing. Galvanized or tinned sheet steel requires occasional painting.

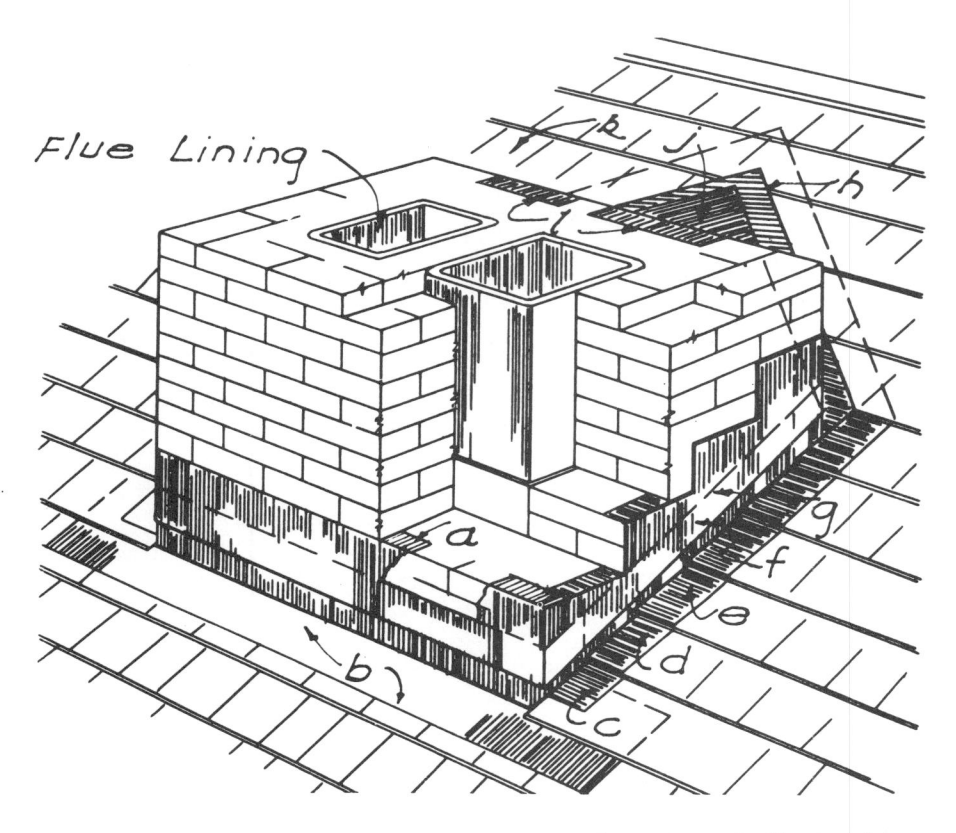

Figure 9. Flashing at a chimney located on the slope of a roof. Sheet metal (h), over cricket (j), extends at least 4 inches under the shingles (k), and is counterflashed at l in joint. Base flashings (b,c,d, and e) and cap flashings (a, f, and g) lap over the base flashings to provide water-tight construction. A full bed of mortar should be provided where cap flashing is inserted in joints.

TOP CONSTRUCTION

Figure 2,A shows a good method of finishing the top of the chimney. The flue lining extends at least 4 inches above the cap or top course of brick and is surrounded by at least 2 inches of cement mortar. The mortar is finished with a straight or concave slope to direct air currents upward at the top of the flue and to drain water from the top of the chimney.

Hoods (Figure 2,C) are used to keep rain out of chimneys and to

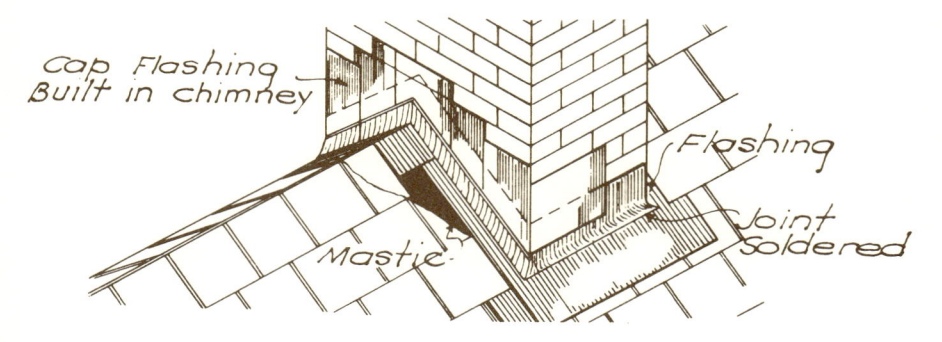

Figure 10. Flashing at a chimney located on a roof ridge.

prevent downdraft due to nearby buildings, trees, or other objects. Common types are the arched brick hood and the flat-stone or cast-concrete cap. If the hood covers more than one flue, it should be divided by wythes so that each flue has a separate section. The area of the hood opening for each flue must be larger than the area of the flue.

Spark arresters (Figure 2,B) are recommended when burning fuels that emit sparks, such as sawdust, or when burning paper or other trash. They may be required when chimneys are on or near combustible roofs, woodland, lumber, or other combustible material. They

Figure 11. Construction of a cricket (j, fig. 9) behind a chimney.

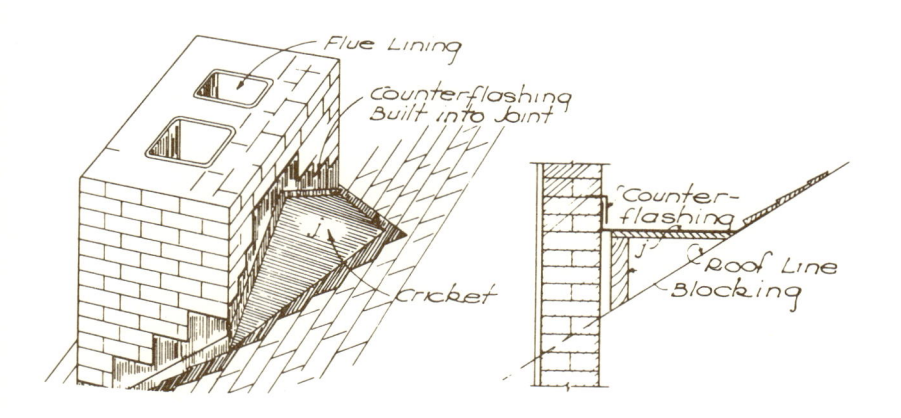

are not recommended when burning soft coal because they may become plugged with soot.

Spark arresters do not entirely eliminate the discharge of sparks, but if properly built and installed, they greatly reduce the hazard. They should be made of rush-resistant material and should have screen openings not larger than five-eighths inch nor smaller than five-sixteenths inch. (Commercially made screens that generally last for several years are available.) They should completely enclose the flue discharge area and must be securely fastened to the top of the chimney. They must be kept adjusted in position and they should be replaced when the screen openings are worn larger than normal size.

INSPECTING THE CHIMNEY FOR LEAKS

The idea of using a chimney for wood fires creates great uneasiness among many homeowners. Chimneys have a reputation for causing major house fires. This is because many homeowners do not know how to inspect a chimney for soundness. The following tests are very effective. However, they must be done carefully. If one feels a half-baked job will result, then rely on a professional.

Take a good look at the outside of the chimney wherever it is visible, both inside and outside the house. Look for evidence of cracks and fissures in brick and mortar. Usually a crack in mortar is indicated by a dark coloring. This is stain left by escaping smoke and heat. Poke at the mortar. If it flakes or crumbles or "chunks," that is the sign of deterioration.

To check the flue for fallen bricks and cracked or broken flue lining, hang an electric light down the chimney on an extension cord. While looking for breakage, check for soot accumulation, too.

If leaks cannot be found using the two previous methods, perform the following smoke test.

Light a small fire in the chimney and watch all exposed areas of the chimney for smoke leaks. If the chimney has a fireplace or a cleanout hole (close to the floor), the fire is relatively easy to manage. If only a stove hole exists, place green grass or leaves in an old-fashioned popcorn popper or similar device with a heavy screen and light the material. Insert it into the stove hole. When smoke begins to issue from the top of the chimney, close up the bottom air hole and then go to the roof and cut off the smoke's escape route using a wet burlap bag. If there are leaks, the smoke will find them, but be sure you perform this test long enough to insure full investigation. Sometimes it takes a while for the pressure within the chimney to build to a point where it forces the smoke through cracks, if any exist.

If the chimney is covered by interior and exterior walls, drill a small hole in several places to determine whether or not smoke leaks from the chimney into the walls.

Needless to say the smoke test should not be undertaken if initial ferreting reveals an unsound chimney.

REPOINTING BRICK CHIMNEYS

Small cracks and leaks can usually be repaired by repointing the bricks. Repointing is the process of removing the loose and crumbling mortar from between the bricks and replacing it with fresh mortar. A trowel, finishing tool and mortar board are the tools necessary for the task. Mortar for chimneys must be specially prepared because it must be able to expand and contract as the chimney is heated and cooled. This mortar consists of one part Portland Cement, one part hydrated lime and six parts of clean sand, measured by volume. It can also be purchased ready-mixed.

First remove the old mortar. Use a cold chisel and dig out the deteriorating mortar until there is a hole or a trench an inch in depth between the bricks. Light tapping on the chisel should suffice.

Second, apply the mortar to the prepared surfaces. Mix it as described above and trowel an easily-handled glob of it onto the mortar board. The trowel is bent so as to have a top side and a bottom. The mortar is carried to the brick on the bottom of the trowel. The bottom is also used to smooth the mortar. Hold the trowel with its edge on the mortar board and scoop up a small elongated ridge of mortar so that it runs the entire length of the trowel. Force this mortar into the prepared holes and cracks. Smooth over first using the bottom of the trowel and then using the point to insure that the mortar fits tightly. Then scrape off the excess using the side of the trowel and return it to the mortar board.

The final step in the process is to use the finishing tool to create the proper smooth edging effect between the bricks. Experience will improve the quality of the finish.

LINING A LINERLESS CHIMNEY

The process that follows will work only on chimneys that are relatively short. The material used is galvanized pipe and if more than one story of a house is to be traversed, the weight of the pipe becomes too much to hold. Before attempting the process, be sure that the chimney is structurally sound. In time the galvanized pipe will rust and it is recommended here as a temporary repair that will last for a good three years, after which it should be replaced.

All the smoke will be carried up the galvanized pipe so it must extend from the stovepipe connection inside the wall up to and above the roof top. All pipe sections are attached with sheet metal screws before they are lowered down to the stove level.

Step one is to enlarge the stove hole where the galvanized pipe will come from the chimney into the room to connect with the stove. Chip away at the brick with a cold chisel and a hammer. The hole should be large enough for the pipe and a guiding hand. Insert a two-by-four through the hole horizontally so that it blocks the chimney. This prevents the pipe from falling too far down the chimney should it slip one's grasp on the roof. One person must remain at the stove hole throughout the process.

Step two is to prepare all materials for ascending the roof. The galvanized pipe sections are about 3 feet long. Two of these sections can be screwed together on the ground. Use an electric drill to make the holes for the sheet metal screws. Other tools that will be needed on the roof are: rope for tying the pipe in place while screw holes are drilled and another section is being fitted over that which already sits in the chimney, and rope for tying a ladder to the chimney.

Measure the distance from the top of the chimney to the stove hole and prepare enough sections for the full descent of the chimney.

The first piece to descend down the chimney is the flexible elbow which will eventually project through the wall to the stove pipe. These elbows can be straightened out for their descent and the person waiting at the bottom can reach through the hole and bend the pipe back to its original form.

When all such preparations have been made, go to the roof. The

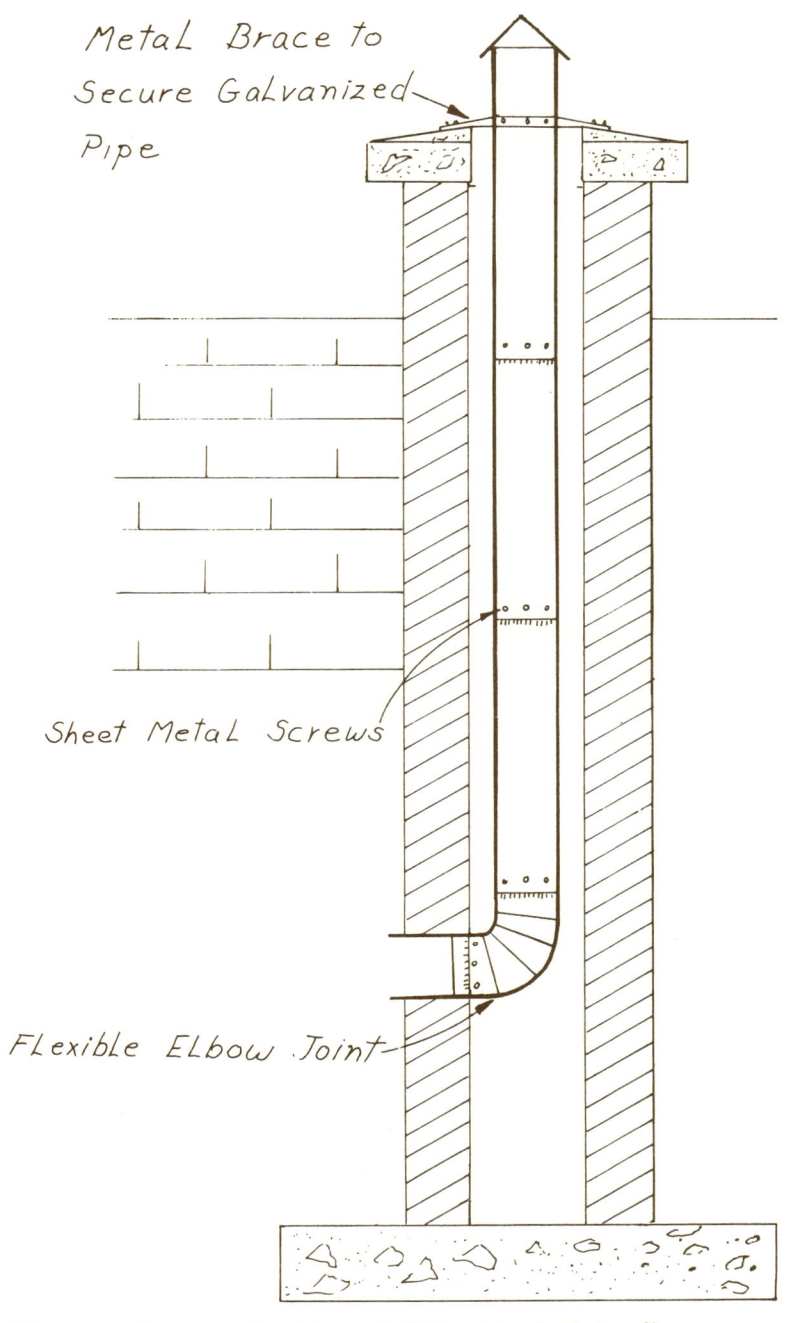

Metal Brace to Secure Galvanized Pipe

Sheet Metal Screws

Flexible Elbow Joint

Figure 12. Crosscut of a chimney holding galvanized pipe liner.

ladder should be secured in place and the pipe sections lined up and readied for use. The extension cord and drill should be secured. The operation begins by lowering the elbow section, attached to a straight section, down the chimney until about a foot of pipe extends out of the top of the chimney. The pipe should then be tied so that it will not slip. Fit the next section over the extruding end and drill holes around it for the sheet metal screws. When the connection is secure, loosen up on the rope slightly and let the pipe slip down until a foot of the new section extrudes from the top of the chimney. Repeat the process until the person stationed below can reach through the hole and grasp the elbow section and direct it through to the room. The upper end of the completed pipe lining should extend above the roof line of the house. A galvanized metal rain cap will add extra life to the finished product.

Finally, fill in the space around the stove hole where the elbow projects into the room. Mortar or fire clay can be used for the job. Hook up the stove and call it a day by the fire.

Another way to repair chimneys which are unlined is to insert tile liners from the top, using a tension catch which will hold the tile liner, allowing it to be lowered from the top. Use sticky cement mortar and lower each tile until it is resting on the top of the previous liner.

The tension catch can be built to release when stress is off or a lead weight can be dropped down the line. You can have an experienced mason advise you about this technique.

PREFABRICATED METAL AND CINDERBLOCK CHIMNEYS

Installing new chimneys has been made easier by the development of prefabricated metal and cinderblock chimneys. The metal versions are more expensive than the cinderblock type, (excluding labor costs) but they are relatively simple to install.

The cinderblock chimney must be built upon a foundation which extends below the ground freeze level. Prefabricated metal chimneys can be hung from the roof and only need to extend to the level of the stove.

To install the metal prefab type, holes are cut in the roof for the chimney and a roof support section is attached. This supports the entire weight of the chimney. The sections are joined together in a simple twist motion. One can shoot them down through floors or run them up a side of the house. The chimneys come in many sizes, and if one is unsure of just how much diameter is needed to vent a stove or fireplace, it is best to err in favor of excess.

The cinderblock chimney starts with the foundation. This should plunge at least six inches beyond the size of the block being used. The depth of the foundation is determined by the freeze line and should go to a depth of at least 18 inches below the line. Reinforcement rods are a good extra to add to any foundation. Clay or sandy soils will indicate the foundation will need a larger footing than if constructed in firm soil. One foot in all directions beyond the base of the chimney should serve for the foundation's boundaries in firm soil.

There is no end to the variety of cinderblock chimneys that can be fashioned. However, those which use one block per level are the simplest to construct. These blocks come with a cast inner circle, or the inner circle can be purchased separately from the outer square. The latter is helpful if the chimney to be constructed must shoot up to unusual heights.

The blocks are secured to one another using mortar. About 3/4's of an inch of mortar on all touching surfaces will do. Lightly tap the block in place and remove excess mortar.

A plumb bob and a level are necessary tools and must be used on each block. The plumb bob is suspended from the top of the roof to insure that the chimney is constructed without leanings. The level should be laid on each new block to insure that it is horizontal.

If the liners are purchased separately from the blocks, lay three blocks and then insert the liner. These must also be mortared together.

No more than five blocks per day should be laid. Too much weight on the fresh mortar will cause it to squeeze out of its joint, thus creating an imbalance to the whole structure. Experienced builders compensate for this by using small wooden supports for the blocks to insure that mortar is allowed to dry properly. However, this is rather much of a skill and it is best for the beginner to stick to the five blocks a day formula.

CLEANING THE CHIMNEY

A major cause of chimney fires can be attributed to soot, which may occur in the form of ashes, tars, or creosote. The latter, if allowed to collect even minimally on the interior walls of a chimney, will one day surely catch fire. It is produced by burning green or partially dried wood. It is also a result of incomplete combustion. These soot deposits can be kept at a minimum by using a product known as Chimney Sweep. However, it will not remove accumulated soot. If used on a weekly basis of 1/2 cup on a working fire and the chimney has been thoroughly cleaned once a year, it will retard the gathering of soot.

To accomplish a major cleaning job, a professional can be hired, but the following tips are very effective and may save some money.

The best equipment is a small Christmas tree and a long rope. Tie the rope to the tree so that two lengths dangle from top and bottom. Feed the length of rope attached to the top of the tree into the chimney from the roof. The person below in the house starts pulling gently on the tree and once the tree has reached the opening in the fireplace, the person on the roof pulls it back up. Repeat this process until soot ceases to fall down the chimney. It is best to acquire the help of a close friend for this task, as the one at the bottom will be rather dirty once the job is accomplished. The same process can be undertaken using a burlap bag filled with newspaper and a few bricks to weigh it down.

One can avoid great collections of soot by learning how to control a fire. Never allow a fire in a stove to "roar." This only creates an excess of draft pressures and will carry the flame into the chimney. Soot deposits will grow swiftly and the licking flames will soon create a chimney fire.

How does one tell if a chimney fire is underway? They make a very obvious roar, accompanied by a "sucking" sound, which represents oxygen being drawn up through the chimney. It is indeed a blast furnace and the heat created will usually set the interior walls and the roof afire. All of that may happen at once. Some brave people like to create a fire in the chimney to burn out a year's collection of soot. Indeed, they are disappointed if one does not occur.

For those of us who would like to know how to put one out, first stop the incoming air at the opening of the fireplace. Pull the stovepipe connection leading to the chimney and stuff the pipe with a wet rag. Second, call the fire department. Third, if at all possible, cover the top of the chimney with wet burlap bags in order to prevent a down draft of oxygen. Some people recommend hosing down the fire from the top of the chimney. If it is a small blaze, maybe. The idea of flames shooting out the top, however, does not make this the best of ideas. If the walls of the house start to smoulder, then perhaps it is the thing to do. A well-kept chimney will rarely present such dangerous choices to its owner.

Around Fireplaces

Fireplaces are not an economical means of heating. And tests indicate that, as ordinarily constructed, they are only about one-third as efficient as a good stove.

However, a well-designed, properly built fireplace can provide additional heat, and all the heat necessary in mild climates. It can enhance the appearance and comfort of the room. It will burn as fuel certain combustible materials that otherwise might be wasted - for example, coke, briquets, and scrap lumber.

A fireplace should harmonize in detail and proportion with the room in which it is located, but safety and utility need not be sacrificed for appearance.

Lord Rumford in the late 1800s wrote a treatise on fireplaces. His research still stands as the definitive statement on fireplace design:

> A fireplace should be no more than one half as deep as it is wide;
> the back wall should have a slope toward the mantel;
> the back wall height should be equal to the width;
> the opening to the flue need not be more than 4 inches wide;
> the chimney should be equipped with a smoke shelf to prevent downdrafts.

This information and some thought about placement of the chimney should be all that is required. Remember that bricks hold heat. If the chimney is in the center of the house, it will radiate heat to all the house, but if it is on an outside wall, it will lose heat to the outside. The center chimney cape with its massive center chimney was an excellent idea.

HEIGHT

Fireplace openings are usually made from 2 to 6 feet wide. The kind of fuel to be burned can suggest a practical width. For example, where cordwood (4 feet long) is cut in half, an opening 30 inches wide is desirable; but where coal is burned, a narrower opening can be used.

Height of the opening can range from 24 inches for an opening 2 feet wide to 40 inches for one that is 6 feet wide. The higher the opening, the more chance of a smoky fireplace.

In general, the wider the opening, the greater the depth. A shallow opening throws out relatively more heat than a deep one, but holds smaller pieces of wood. You have the choice, therefore, between a deeper opening that holds larger, longer-burning logs and a shallower one that takes smaller pieces of wood but throws out more heat. In small fireplaces, a depth of 12 inches may permit good draft, but a minimum depth of 16 inches is recommended to lessen the danger of brands falling out on the floor. Suitable screens should be placed in front of all fireplaces to minimize the danger from brands and sparks.

Second-floor fireplaces are usually made smaller than first-floor ones, because of the reduced flue height.

FOOTINGS AND HEARTH

Foundation-and-footing construction of chimneys with fireplaces is similar to that for chimneys without fireplaces. Be sure the footings rest on good firm soil below the frostline.

The fireplace hearth should be made of brick, stone and terra cotta, or reinforced concrete at least 4 inches thick. It should project at least 20 inches from the chimney breast and should be 24 inches wider than the fireplace opening (12 inches on each side).

The hearth can be flush with the floor so that sweepings can be brushed into the fireplace or it can be raised. Raising the hearth to various levels and extending in length as desired is presently common practice, especially in contemporary design. If there is a basement, a convenient ash dump can be built under the back of the hearth. (Figure 2)

In buildings with wooden floors, the hearth in front of the fireplace should be supported by masonry trimmer arches or other fire-resistant construction. Wood centering under the arches used during construction should be removed when construction is completed (Figure 1).

WALLS, JAMBS, AND LINTEL

Building codes generally require that the back and sides of fireplaces be constructed of solid masonry or reinforced concrete at least 8 inches thick and be lined with firebrick or other approved noncombustible material not less than 2 inches thick or steel lining not less than one-fourth inch thick. Such lining may be omitted when the walls are of solid masonry or reinforced concrete at least 12 inches thick.

The jambs of the fireplace should be wide enough to provide stability and to present a pleasing appearance. Often they are faced with ornamental brick or tile.

For a fireplace opening 3 feet wide or less, the jambs can be 12 inches wide if a wood mantel will be used or 16 inches wide if they will be of exposed masonry. For wider fireplace openings, or if the fireplace is in a large room, the jambs should be proportionately wider.

Fireplace jambs are frequently faced with ornamental brick or tile.

No woodwork should be placed within 6 inches of the fireplace opening. Woodwork above and projecting more than 1½ inches from the fireplace opening should be placed not less than 12 inches from the top of the fireplace opening.

A lintel must be installed across the top of the fireplace opening to support the masonry.

For fireplace openings 4 feet wide or less, ½- by 3-inch flat steel bars, 3½- by ¼-inch angle irons, or specially designed damper frames may be used. Wider openings will require heavier lintels.

If a masonry arch is used over the opening, the fireplace jambs must be heavy enough to resist the thrust of the arch.

Recommended dimensions for fireplaces and size of flue lining required

Letters at the head of each column refer to Figure One

Size of fireplace opening		Depth	Minimum width of back wall	Height of vertical back wall	Height of inclined back wall	Size of flue lining required	
Width	Height					Standard rectangular (outside dimensions)	Standard round (inside diameter)
w	h	d	c	a	b		
Inches	Inches	Inches	Inches	Inches	Inches	Inches	Inches
24	24	16-18	14	14	16	8½x13	10
28	24	16-18	14	14	16	8½x13	10
30	28-30	16-18	16	14	18	8½x13	10
36	28-30	16-18	22	14	18	8½x13	12
42	28-32	16-18	28	14	18	13x13	12
48	32	18-20	32	14	24	13x13	15
54	36	18-20	36	14	28	13x18	15
60	36	18-20	44	14	28	13x18	15
54	40	20-22	36	17	29	13x18	15
60	40	20-22	42	17	30	18x18	18
66	40	20-22	44	17	30	18x18	18
72	40	22-28	51	17	30	18x18	18

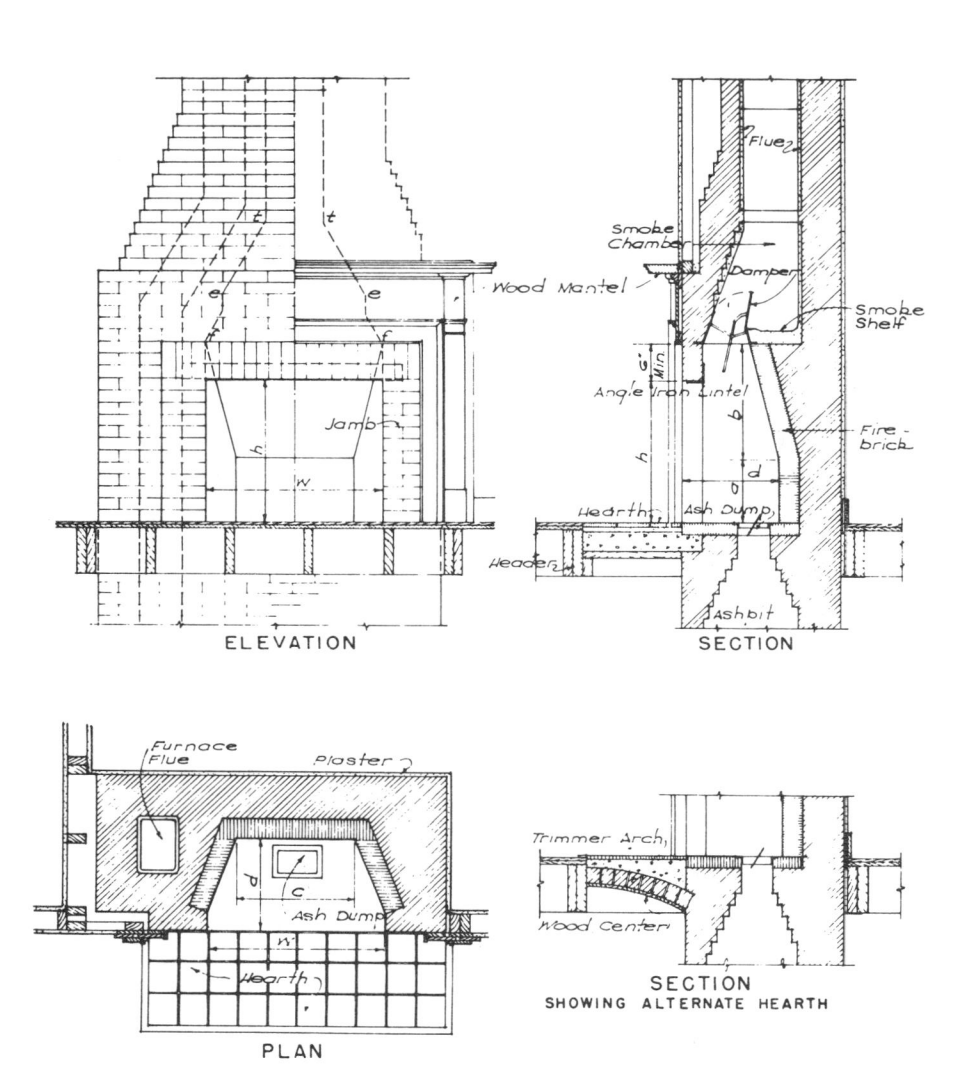

ELEVATION

SECTION

PLAN

SECTION
SHOWING ALTERNATE HEARTH

Figure 1. Construction details of a typical fireplace. (The letters indicate specific features discussed in the text). The lower right-hand drawing shows an alternate method of supporting the hearth.

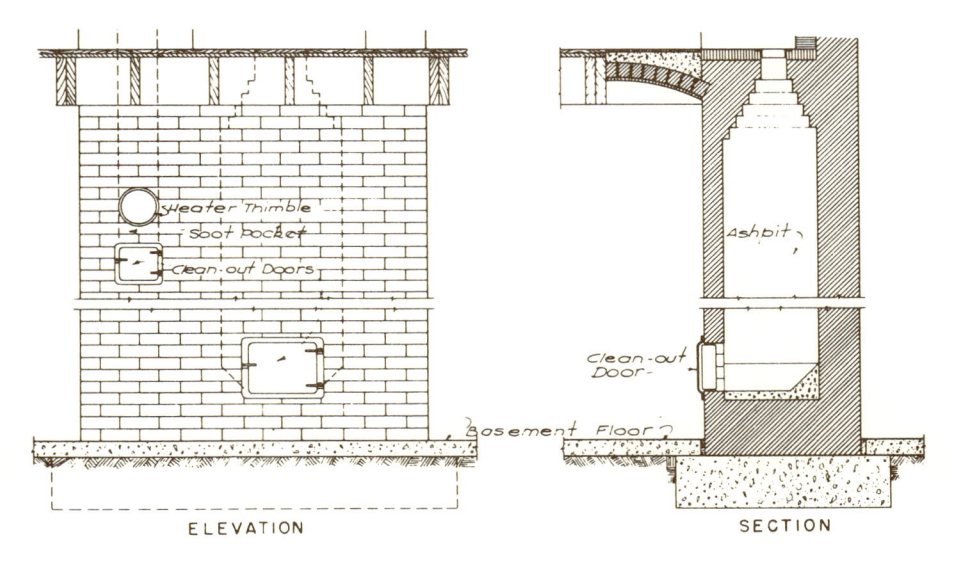

ELEVATION   SECTION

Figure 2.   An ashpit for a fireplace should be of tight masonry and should be provided with a tightly fitting iron cleanout door and frame 10 inches high and 12 inches wide.   The left-hand drawing also shows a cleanout for a furnace flue.

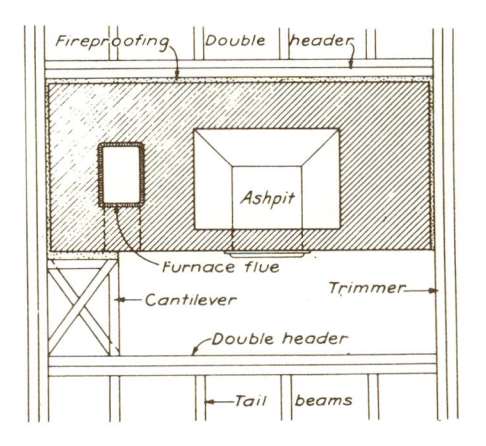

Figure 3.   Installation of floor framing around a chimney and hearth.   Where a header is more header is more than 4 feet long, it should be doubled as shown.   If it supports more than four tail beams, its ends should be supported in metal joist hangers.   The framing may be placed one-half inch from masonry chimney walls 8 inches thick.

THROAT

Proper construction of the throat area is essential for a satisfactory fireplace. (ff, Figure 1)

The sides of the fireplace must be vertical up to the throat, which should be 6 to 8 inches or more above the bottom of the lintel.

Area of the throat must be not less than that of the flue -- length must be equal to the width of the fireplace opening and width will depend on the width of the damper frame (if a damper is installed).

Five inches above the throat the sidewalls should start sloping inward to meet the flue. (at tt, Figure 1)

DAMPER

A damper consists of a cast-iron frame with a hinged lid that opens or closes to vary the throat opening.

Dampers are not always installed, but they are definitely recommended, especially in cold climates.

With a well-designed, properly installed damper, one can:

Regulate the draft.

Close the flue to prevent loss of heat from the room when there is no fire in the fireplace.

Adjust the throat opening according to the type of fire to reduce loss of heat. For example, a roaring pine fire may require a full throat opening, but a slow-burning hardwood log fire may require an opening of only 1 or 2 inches. Closing the damper to that opening will reduce loss of heat up the chimney.

Close or partially close the flue to prevent loss of heat from the main heating system. When air heated by a furnace goes up a chimney, an excessive amount of fuel may be wasted.

Close the flue in the summer to prevent insects from entering the house through the chimney.

Dampers of various designs are on the market. Some are designed to support the masonry over fireplace openings, thus replacing ordinary lintels.

Responsible manufacturers of fireplace equipment usually offer assistance in selecting a suitable damper for a given-size fireplace. It is important that the full damper opening equal the area of the flue.

SMOKE SHELF AND CHAMBER

A smoke shelf (Figure 1) prevents downdraft. It is made by setting the brickwork at the top of the throat back to the line of the flue wall for the full length of the throat. Depth of the shelf may be 6 to 12 inches or more, depending on the depth of the fireplace.

The smoke chamber is the area from the top of the throat (ee, Figure 1) to the bottom of the flue (tt, Figure 1). As indicated under "Throat," the sidewalls should slope inward to meet the flue.

The smoke shelf and the smoke-chamber walls should be plastered with cement at least one-half inch thick.

FLUE

Proper proportion between the size (area) of the fireplace opening, size (area) of the flue, and height of the flue is essential for satisfactory operation of the fireplace.

The area of a lined flue 22 feet high should be at least one-twelfth of the area of the fireplace opening. The area of an unlined flue or a flue less than 22 feet high should be one-tenth the area of the fireplace opening.

The table preceding Figure 1 lists dimensions of fireplace openings and, in the last two columns, indicates the size of the flue lining required. From this table you can determine the size of lining required for a given-size fireplace opening and also the size of opening to use with an existing flue.

MODIFIED FIREPLACES

Modified fireplaces are manufactured fireplace units, made of heavy metal and designed to be set in place and concealed by the usual brickwork or other construction. They contain all the essential fireplace parts--firebox, damper, throat, and smoke shelf and chamber. In the completed installation, only grilles show.

Figure 4 shows one of the several designs of modified fireplaces available.

Modified fireplaces offer two advantages:

The correctly designed and proportioned firebox provides a ready-made form for the masonry, which reduces the chance of faulty construction and assures a smokeless fireplace.

When properly installed, the better designed units heat more efficiently than ordinary fireplaces.

They circulate heat into the cold corners of rooms and can deliver heated air through ducts to upper adjoining rooms.

The use of a modified fireplace unit can increase the cost of a fireplace (although manufacturers claim that labor, materials, and fuel saved offset any additional cost). It should not be necessary to use one merely to insure an attractive, well-proportioned fireplace; you can build an equally attractive and satisfactory masonry fireplace by careful construction.

Even a well-designed modified fireplace unit will not operate properly if the chimney is inadequate. Therefore, proper chimney construction is as important for these units as it is for ordinary fireplaces.

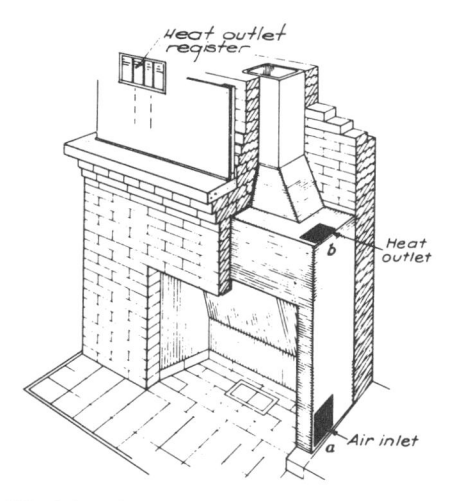

Figure 4. A modified fireplace. Air is drawn through inlet "a" from the room being heated. It is heated by contact with the metal sides and back of the fireplace, rises by natural circulation, and is discharged through outlet "b" The inlets and outlets are connected to registers which may be located at the front (as shown) or on the wall of an adjacent or second-story room, or at the ends of the fireplace.

SOME STOVE MANUFACTURERS

Portland Franklin Stove Foundry Co.
Box 1156
Portland, Maine 04100
    Space Heaters, Franklin Stoves, Wood Cook Stoves, Free-
    Standing Fireplaces.

King Stove and Range Co.
Box 730
Sheffield, Alabama 35660
    Space Heaters, Franklin Stoves, Wood Cook Stoves, Automatic
    Circulating Heaters.

Birmingham Stove and Range Co.
Box 3593
Birmingham, Alabama 35202
    Space Heaters, Franklin Stoves, Wood Cook Stoves, Automatic
    Circulating Heaters.

Brown Stove Works, Inc.
Cleveland,
Tennessee 37311
    Space Heaters.

U. S. Stove Co.
South Pittsburg,
Tennessee 37380
    Space Heaters, Franklin Stoves, Automatic Circulating Heaters.

Riteway Manufacturing Co.
Box 6
Harrisonburg, Virginia 22801
    Wood-burning central heating systems and Space Heaters.

Shenandoah Manufacturing Co., Inc.
Box 839
Harrisonburg, Virginia 22801

Ashley Automatic Heater Co.
1604 Seventeenth Avenue, S.W.
Sheffield, Alabama 35660
    Automatic Circulating Heaters

Scandinavian
    Jøtul — Kristia Associates, Portland, Maine — Importers
    Trolla — Portland Stove Foundry Co., Portland, Maine

Austrian
    Styria —   Distributed by Merry Music Box, 18 High Street,
             Wiscasset, Maine
             Space Heaters, Cooking Ranges, Automatic Circulating
             Heaters.

Canadian
    Lunenburg Foundry, Lunenburg, Nova Scotia
    Fawcett Limited, Sackville, New Brunswick

ACKNOWLEDGMENTS

The author wishes to gratefully acknowledge the following people for their advice and assistance in gathering portions of the text:

The Portland Stove Foundry, Portland, Maine, for the use of the wood stove engravings found in Chapter Three.
Eva Horton of Kristia Associates, Portland, Maine, importers of Norwegian stoves for her advice on the same.
The Saugus Iron Works of Saugus, Massachusetts, for their help in gathering historical materials.
Peg Doggett for her valuable assistance on the material found in Chapter Four.
And for the advice and assistance rendered by friends Peter Baade and Joe Steinberger.
And last, but certainly not least, Bonnie Grimm for her assistance with illustrations.